Fiscal Systems for Hydrocarbons

Design Issues

Silvana Tordo

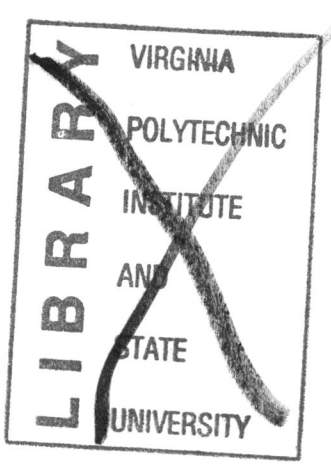

THE WORLD BANK
Washington, D.C.

World Bank Working Papers are published to communicate the results of the Bank's work to the development community with the least possible delay. The manuscript of this paper therefore has not been prepared in accordance with the procedures appropriate to formally-edited texts. Some sources cited in this paper may be informal documents that are not readily available.

ISBN-13: 978-0-8213-7266-1
eISBN: 978-0-8213-7267-8
ISSN: 1726-5878   DOI: 10.1596/978-0-8213-7266-1

Silvana Tordo is Senior Energy Economist in the Oil, Gas, and Mining Policy Division of the World Bank.

Library of Congress Cataloging-in-Publication Data has been requested.

Contents

LIST OF FIGURES

LIST OF GRAPHS

Abstract

Although host governments and the investors may share one common objective—the desire for the project to generate high levels of revenue—their other goals are not entirely aligned. Host governments aim to maximize the rent for their country over time, while achieving other development and socioeconomic objectives. Investors' aim is to ensure that the return on investment is consistent with the risk associated with the project, and with their corporations' strategic objectives. To reconcile these often conflicting objectives, more and more countries rely on transparent institutional arrangements and flexible, neutral fiscal regimes. This paper examines the key elements of the legal and fiscal frameworks utilized in the petroleum sector and aims to outline desirable features that should be considered in the design of fiscal policy with the objective of optimizing the host government's benefits, taking into account the effect that this would have on the private sector's investment.

Acknowledgments

The author thanks Daniel Johnston, Charles P. McPherson, Hossein Razavi, Robert W. Bacon, and Peter van der Veen for their constructive comments, and Randy Hecht for her helpful edits and suggestions.

CHAPTER 1

Introduction

The global market for oil and gas exploration has evolved to the point that much of the world's surface open to exploitation has taken on some of the characteristics of a commodity. Governments compete for capital and technology to develop their hydrocarbon sector. In order to devise and apply the appropriate policies, strategies and tactics, each must assess its position in the global marketplace and evaluate its particular situation, boundary conditions, concerns and objectives. Companies look for investment opportunities that suit their corporate strategies and risk-reward profiles. The initial decision to invest and the resulting allocation of revenue and benefits are greatly influenced by the content of existing legal arrangements and fiscal policies.

The fiscal regime can be used to convert a government's policy into economic signals to the market, and influence investment decisions, provided that the framework is clear, is not changed retroactively, and does not discriminate among the actors. Several countries have used favorable taxation of oil and gas to support the development of the sector in addition to relevant sector reforms. The challenge of an efficient fiscal system is to induce maximum effort from the oil companies while ensuring that the host government is adequately compensated.

In designing a fiscal system, a government has to answer the following questions: What is the effect of the fiscal regime on oil/gas output? Does it discourage the development of marginal fields? Does it influence the pace of development? Does it favor early abandonment? Is it insensitive to oil/gas price and cost variation? In other words, how flexible, neutral and stable is the fiscal regime?

Many fiscal systems around the world make use of sliding scales for the determination of at least one of the following parameters: royalty, bonuses, profit oil/gas split, cost recovery, and taxes. Sliding scales introduce flexibility into the system by allowing it to respond to changes in project variables. Unfortunately, the vast majority of these sliding scales are linked to daily or cumulative production targets. Hence they are insensitive to changes in economic variables. No wonder that the persistently high level of oil prices in recent years has pushed many host governments to seek improvement in their contractual terms.

High oil prices have also triggered higher demand for services and equipment, which in turn has increased their cost.[1] As many fiscal systems[2] were designed when oil prices were in the US$15-18 barrel range and finding and development costs were US$5-9 barrel, these systems no longer efficiently capture the projects' economic rent.

High risks and long project cycles are key elements of the oil and gas industry. As risks can differ substantially by project and over time, an efficient fiscal system needs to be flexible enough to allocate risks equitably, thus minimizing the need for and cost of negotiations or renegotiations. Such a system would be, at least in theory, more stable and better suited to mitigating the investment risk. If correctly designed, the fiscal system has the potential to reduce the procyclicity of investment: a less variable flow of investment is more likely to support the creation of spare capacity, thus reducing price volatility.

In today's competitive market, many diverging interests must be recognized and accommodated to establish an effective and attractive legal and fiscal framework for hydrocarbon exploration and production. No ideal or model regime is available for policy makers to adopt. Each country's circumstances, needs, and objectives define the key features of an appropriate legal and fiscal framework. This paper provides an overview of the key features of petroleum fiscal systems around the world and attempts to outline desirable features for designing a fiscal regime for the management of a country's petroleum endowment. Chapters 2 and 3 provide background material on, respectively, the stages of an oil and gas project and the type of legal arrangements normally used in the petroleum sector. The relative advantages and disadvantages of the tax and non-tax instruments used in petroleum fiscal regimes are discussed in Chapter 4. Chapter 5 outlines the features of successful fiscal regimes, while system measures and economic indicators are described in Chapter 6. Finally, in Chapter 7, a sensitivity analysis is used to illustrate some typical fiscal systems' design issues.

---

1. The World Offshore Oil and Gas Products and Spend Forecast 2007-11, published by Douglas-Westwood, predicts this trend to continue over the next five years, with particular emphasis on deep water floating and subsea production solutions. Operating costs are also expected to increase by more than 50 percent by 2011 as a result of increasing output and producing a higher share of more expensive oil. The impact will differ among regions.

2. Including some R-Factor and RoR-based systems. For a definition of R-Factor and RoR see Chapter 4.

CHAPTER 2

The Life Cycle of a Petroleum Project

The stages of a typical oil and gas project can be described as follows:

1. *Licensing:* In most cases the host government grants a license (lease, or block area) or enters into a contractual arrangement with an oil company or group of oil companies to explore for and develop a field without transferring the ownership of the mineral resources.

2. *Exploration:* After acquiring the rights, the oil company carries out geological and geophysical surveys such as seismic surveys and core borings. The data so acquired are processed and interpreted and, if a play appears promising, exploratory drilling is carried out. Depending on the location of the well a drilling rig, drill ship, semi-submersible, jack-up, or floating vessel will be used.

3. *Appraisal:* If hydrocarbons are discovered, further delineation wells are drilled to establish the amount of recoverable oil, production mechanism, and structure type. Development planning and feasibility studies are performed, and the preliminary development plan is used to estimate the development costs.

4. *Development:* If the appraisal wells are favorable and the decision is made to proceed, then the next stage of development planning commences using site-specific geotechnical and environmental data. Once the design plan has been selected and approved, contractors are invited to bid for tender. Normally, after approval of the environmental impact assessment by the relevant government entity, development drilling is carried out and the necessary production and transportation facilities are built.

3

5. *Production:* Once the wells are completed and the facilities are commissioned, production starts. Workovers[3] must be carried out periodically to ensure the continued productivity of the wells, and secondary and/or tertiary recovery[4] may be used to enhance productivity at a later time.

6. *Abandonment:* At the end of the useful life of the field, which for most structures occurs when the production cost of the facility is equal to the production revenue (the so-called "economic limit"), a decision is made to abandon. For a successful removal, operators generally begin planning one or two years prior to the planned date of decommissioning (or earlier depending on the complexity of the operation).

Figure 1 provides a graphic representation of the project cycle.

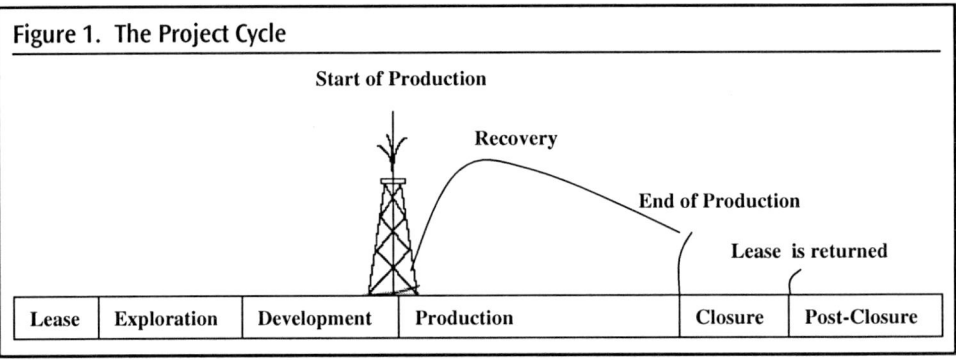

Figure 1. The Project Cycle

The risk profile of the project changes during its life cycle. Risks can be grouped under three main categories: geological, financial, and political. In general terms, while geological risk begins to diminish after a discovery, the political and financial risks intensify. One of the reasons for this is that the bargaining power and relative strength of the investors' and the host government's positions shift during the cycle of petroleum exploration and development. By the time production commences, capital investment is a sunk cost, and facilities installed in foreign countries represent a source of vulnerability to the investor.

---

3. Any operation performed on a well subsequent to its completion.

4. In the first stage of hydrocarbon production natural reservoir energy, such as gasdrive, waterdrive or gravity drainage, displaces hydrocarbons from the reservoir into the wellbore and up to surface. Initially, the reservoir pressure is considerably higher than the bottomhole pressure inside the wellbore. This high natural differential pressure drives hydrocarbons toward the well and up to surface. However, as the reservoir pressure declines because of production, so does the differential pressure. When the reservoir pressure is so low that the production rates are not economical, or when the proportions of gas or water in the production stream are too high, secondary or tertiary recovery methods may be used. Secondary recovery consists of injecting an external fluid, such as water or gas, into the reservoir through injection wells located in rock that has fluid communication with production wells. The purpose of secondary recovery is to maintain reservoir pressure and to displace hydrocarbons toward the wellbore. Tertiary recovery (or enhanced oil recovery) involves the use of sophisticated techniques that alter the original properties of the oil. Enhanced oil recovery can begin after a secondary recovery process or at any time during the productive life of an oil reservoir. Its purpose is not only to restore formation pressure, but also to improve oil displacement or fluid flow in the reservoir.

Although many of the variables that affect the profitability of a petroleum project are beyond the control of both the host government and the investing companies, the host government can take actions to minimize uncertainty. Options include providing potential investors with access to existing geological and geophysical data; strengthening macroeconomic and fiscal stability; improving transparency and the rule of law; promoting contract stability; and signing/ratifying relevant international conventions.

Project uncertainty correlates directly with the cost of the investment: reducing uncertainty results in a reduction of the cost of capital, which in turn increases the rent potentially available for taxation. Risk management is a key feature of the oil industry. Companies hedge against risk by investing in a diverse portfolio of projects and by involving multiple partners. Countries may not have the same ability to diversify their investments. Hence they hedge against risk by establishing flexible fiscal systems[5] and transferring part of the risk to oil companies.

---

5. See Chapter 5 below.

CHAPTER 3

Legal Arrangements in the Petroleum Industry

The legal basis for hydrocarbon exploration, development and production is normally established in a country's constitution.[6] Normally, the hydrocarbon law, formulated at parliamentary level, sets out the principles of law, while those provisions that do not affect principles of law, or that may need periodic adjustments (such as technical requirements, administrative procedures, and administrative fees), are set in regulations.[7] Governments grant exploration, development and production rights in particular areas or blocks by means of concessions or contracts, depending on their legal systems. Where no hydrocarbon law exists, comprehensive contractual agreements between host governments and investors are used.[8]

Various legal systems have been developed to address the rights and obligations of host government and of private investors. These can be grouped under two families: concessionary systems and contractual systems (see Figure 2).

---

6. The consistency of the legal framework with the constitutional foundation affects the security and stability of the legal framework. This issue is particularly significant because many countries' constitutions differ substantially in the degree to which they recognize or guarantee private property rights or prohibit private parties or foreigners from acquiring property rights in general and mineral rights in particular; vest the authority to grant petroleum rights in the state or provincial governments or agencies rather than the national government, vest the authority to regulate specific matters in special agencies (i.e., environment protection) or in the executive branch (for example, taxation, foreign exchange, employment, and so on) or in the judiciary (settlement of disputes). Due to the capital intensive and long term nature of petroleum projects, certainty of rights is particularly important for private investors.

7. These are normally issued at the executive or ministerial level and do not require the legislative branch's approval.

8. This approach may be favored by those countries that face the uncertainty of entering the sector for the first time or in cases where the importance of the petroleum activity may not justify the design of unique policy regimes.

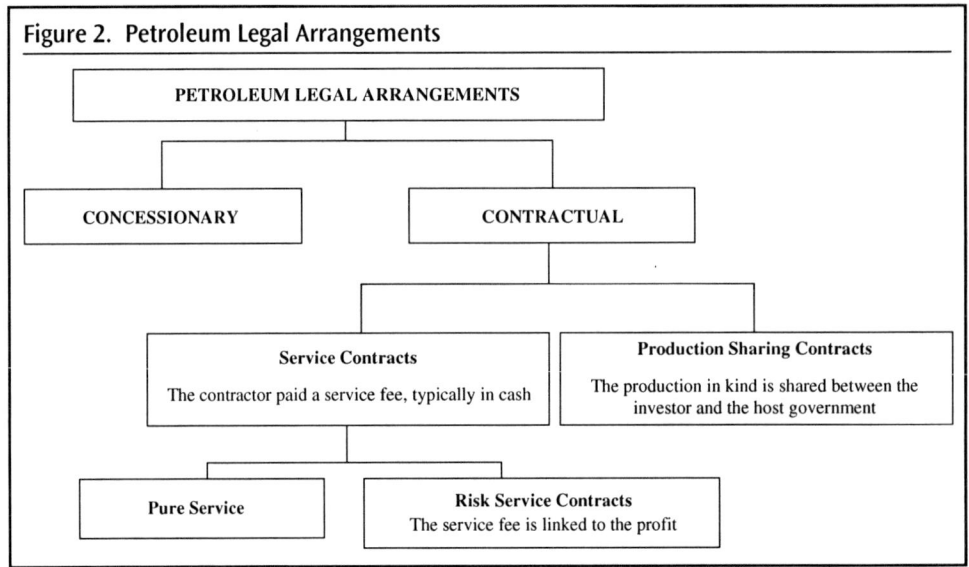

Figure 2. Petroleum Legal Arrangements

Source: Adapted from Johnston (1994b).

In both systems, the investor assumes all risks and costs associated with hydrocarbon exploration, development and production, and receives compensation adequate to the risk. Normally, the investment risks are assumed by oil companies rather than the state/owner of the resource. In general terms, the higher the risk of investment activities in a country, the higher the portion of the rent received by the investor.[9]

The fundamental difference between concessionary and contractual systems relates to the ownership of the natural resources:

- Under a concessionary system, the title to hydrocarbons passes to the investor at the borehole. The state receives royalties and taxes in compensation for the use of the resource by the investor. Title to and ownership of equipment and installation permanently affixed to the ground and/or destined for exploration and production of hydrocarbons generally passes to the state at the expiry, or termination, of the concession (whichever is earlier). The investor is typically responsible for abandonment.
- Under a contractual system, the investor acquires the ownership of its share of production only at the delivery point.[10] Title to and ownership of equipment and installation permanently affixed to the ground and/or destined for exploration and production of hydrocarbons generally passes to the state immediately. Furthermore, unless specific provisions have been included in the contract (or in the relevant legislation) the government (or the national oil company, "NOC") is typically legally responsible for abandonment.

---

9. Although historical considerations influence the definition of "adequate compensation," project specific elements and future expectations are also important.

10. Under a service contract, the contractor never acquires the title to the resource. On the contrary, he is paid a fixed or variable fee for his services. In some service contracts the fee is paid in kind. Except for the nature of the payment Production Sharing Contracts and Risk Service Contracts are very similar.

Table 1. Key Features of Concessionary and Contractual Systems	
Concessionary Systems	**Contractual Systems**
• In its most basic form, a concessionary system has three components: royalty; deductions (such as operating costs, depreciation, depletion and amortization, intangible drilling costs); and tax.	• Under a production sharing contract (PSC) the contractor receives a share of production for services performed. In its most basic form, a PSC has four components: royalty, cost recovery, profit oil, and tax.
• The royalty is normally a percentage of the proceeds of the sale of hydrocarbon[11]. It can be determined on a sliding scale, the terms of which may be negotiable or biddable, and paid in cash or in kind. The royalty represents a cost of doing business and is thus tax-deductible.	• Similar to concessionary systems. In addition, normally royalties are not cost recoverable.
• The definition of fiscal costs is described in the legislation of the country or in the particular concession agreement. Royalties and operating expenditures are normally expensed in the year in which they occur, and depreciation is calculated according to applicable legislation. [12] Some countries allow the deduction of investment credits, interest on financing, and bonuses.	• Fiscal costs are defined and rules for amortization and depreciation are established in the legislation of the country or in the particular PSC. After payment of royalties, the contractor is allowed to recover costs in accordance with contractual provisions (a cost recovery limit may apply). The remainder of the production is split between the host government and the oil company at a stipulated (often negotiated) rate.
• The taxable income under a concessionary agreement may be taxed at the country's basic corporate tax rate. Special investment incentive programs and special resource taxes may also apply. Tax losses are normally carried forward until full recovery. [13]	• Corporate taxes may apply or may be paid by the host government or its NOC on behalf of the contractor. Income tax is calculated on taxable income (revenue net of royalties, allowable costs, and government share of profit oil). Tax losses are normally carried forward until full recovery.[13] In most countries, when cost recovery limits exist, the company's share of profit oil in any given accounting period is not the taxable base[14].

---

11. In some cases the royalty is calculated on net production. Some countries use fiscal prices for the purpose of royalty and corporate tax calculation. These prices are defined periodically and are normally linked to international market prices. The majority of the countries refer to arms length sales to third parties. Whether or not a country uses fiscal prices, deductions or additions are normally allowed to take into account differences in quality between the reference crude (gas) and the particular crude (gas) as well as transport costs.

12. The exact manner in which costs are capitalized or expensed depends on the tax regime of the country and the manner in which rules for integrated and independent producers vary. The successful-efforts and full-cost methods used in oil and gas accounting are discussed in detail in Gallun et al., 2001. In general terms, if costs are capitalized, they may be expensed through the statutory amortization and depreciation schedule, through abandonment, impairment, or depletion. If they are expensed, they are treated as period expenses and charged against revenue in the current period. The primary difference between the two methods is the timing of the expense against revenue and the manner in which costs are accumulated and amortized.

13. Several countries limit the number of years for tax loss carry forward.

14. In fact, the company may receive a share of profit oil but may not be in a taxable position.

Given the risky nature of the industry, in both types of legal systems the investor's ability to share the risk by transferring all or part of its rights to other investors, and the objectivity and transparency of the conditions for government approval or denial of such transfer (including any relevant performance guarantee) are an important element of the overall attractiveness of a country's regime.

The key features of concessionary and contractual systems are summarized in Table 1. Table 2 outlines the main differences between concessions and production sharing contracts.

Table 2. Main Differences between Concessionary Systems and Production Sharing Contracts		
	Concessionary Systems	Production Sharing Contracts
Ownership of nation's mineral resources	Held by sovereign state	Held by sovereign state
Title transfer point	At the wellhead	At the export point
Company entitlement	Gross production less royalty	Cost oil/gas + profit oil/gas
Entitlement percentage	Typically 90%	Typically 50–60%
Ownership of facilities	Held by company	Held by the state
Management and control	Typically less government control	More direct government control and participation
Government participation (carried working interest)	Less likely	More likely
Ring fencing	Less likely	More likely

Source: Johnston (1994b).

Although in most countries, all matters related to petroleum exploration, development and production tend to be governed by sector specific legislation and regulation, countries that have recently reformed their hydrocarbon sector have shown a preference for the establishment of modular legal frameworks. In these cases, all matters related to hydrocarbon rights and their use are governed by the hydrocarbon law/regulations; all matters relating to taxation are defined in the tax code/regulations; all issues relating to environment protection are defined in the environmental law/regulations; and so on. Thus, the hydrocarbon law incorporates other laws by reference. Modularity increases transparency and accountability, reduces administration costs, and facilitates compliance.[15] The topics typically addressed in modern legal frameworks are summarized in Appendix A.

---

15. A clear, simple and non-discretionary legal and regulatory framework is an important factor for attracting foreign investment. This affects the entire value chain from the award of exploration and production rights to the disclosure of information that affects the citizenry. There are various ways of improving the transparency in the management and oversight of the sector: the standardization of the terms of exploration and production, the reduction of the discretion of the administrative authorities, the simplification of awarding and permitting procedures, the development of an efficient and functioning open title system, the adoption of standardized form of agreements, the predefinition of standard shape form of blocks, the granting of greater operating freedom to the contractors, the adherence to international arbitration (in particular where the local court system does not provide sufficient guarantees), and the respect of international disclosure practice are examples in this direction.

CHAPTER 4

Fiscal Regimes for the Petroleum Sector

Tax and Non-Tax Instruments

Petroleum activities around the world are subject to a great variety of taxation instruments. These include taxes that apply to all other sectors of the economy and taxes that are specific to the oil industry. In addition, non-tax forms of rent collection (like surface fees, bonuses and production sharing) are typically used.

Special provisions are often included in petroleum fiscal regimes to modify the timing or magnitude of the revenue appropriations. These provisions are normally intended as incentives designed to attract investors, or to accommodate unique attributes of a petroleum asset, or to influence the choices of the investors toward specific public policy goals.

The most commonly used special provisions include:

Accelerated capital cost allowances	Assets are depreciated in many ways over their expected life (useful life of equipment, economic life of the reservoir). The methods used in the industry are: (a) straight-line (equal annual deductions); (b) declining balance (straight-line depreciation calculated for the remaining value of the asset each year); (c) double declining balance (doubles straight-line depreciation for the remaining value the asset each year); (d) sum of year digits (based on an inverted scale that is the ratio of the number of digits in a given year divided by the total of all years digits); and (e) unit of production (the capital cost of equipment, after deduction of the accumulated depreciation and of the salvage value, is multiplied by the ratio between the total production in a year and the recoverable reserves remaining at the beginning of the tax year).
Depletion allowances	The depletion allowance is the deduction from gross income allowed to investors in exhaustible commodities (such as minerals, oil, or gas) for the depletion of the deposits. The theory behind the allowance is that an incentive is necessary to stimulate investment in this high-risk industry: as the reservoir depletes, the company will need to undertake more exploration to find new reservoirs. The depletion allowance

	is meant to subsidize further exploration. Since the industry is a global one, it is quite likely that the depletion allowance may be used to subsidize exploration in competing countries. For this among other reasons, depletion allowances are granted/have been granted by only a few countries: Barbados, Canada, Pakistan and the USA. The Filipino Participation Incentive Allowance—FPIA - is similar to a depletion allowance.
Interest deduction rules	Project financing is quite common for large projects or for small oil companies. Normally interests on loans are deductible from taxable income and qualify for cost recovery. Inter-company interests may also be cost recoverable and tax deductible, if calculated on an arms-length basis.
Loss carry forward	This refers to the ability of a company to "carry forward" losses from one year to offset tax liability in future years. When limitations apply the loss can be carried forward for a set number of years (normally 5 to 7) after which the benefit expires. In most cases, unlimited loss carry forward is granted.
Investment credits	In some countries, governments provide an incentive to investors by allowing them to recover an additional percentage of tangible capital expenditure (also known as investment uplifts or "allowances" and investment credits). In some cases investment credits can be taxable.
Tax holidays	When capital investment in a project is considerable, the host government may grant tax holidays to investors. For example, Myanmar offers a three year tax holiday period on income tax in PSC (www.energy.gov.mm/MOGE_3.htm). Tax holidays provide a valuable advantage to investing companies that can accelerate the project payback. On the other hand, host governments should be careful in utilizing this mechanism to attract investors.
Stability provisions	See Chapter 5.

A variety of costs are also imposed on companies that affect the profitability of their operations. Some are fairly common while other reflect specific countries' conditions. These costs include inter-company services, valuation of oil and gas, foreign exchange regulations, domestic market obligations, government equity, performance bonds, land owner compensations, local content obligations, and requirement intended to ensure good environmental practices and adequate site reclamation funding.

A description of the main tax and non tax instruments commonly used in the petroleum industry is given in Appendix B, Tables 7 to 19.

CHAPTER 5

Designing Efficient Fiscal Systems

A lthough the host government and the investor may share one common objective—the desire for the project to generate high levels of revenue their other objectives are not entirely aligned:

■ Host governments aim to obtain the maximum value (not volume) for their countries over time in terms of net receipts for treasury. Their goal is to maximize the wealth from their natural resources and, at the same time, attract foreign investment. Host governments also have development and socioeconomic objectives, such as job creation, transfer of technology, and development of local infrastructure.

■ Oil companies aim to ensure that the return on capital is consistent with the risk associated with the project and with the strategic objectives of the corporation.

From the government's standpoint, this means the design of a tax system that:

  (i) supports macroeconomic stability by providing predictable and stable tax revenue flows;

 (ii) permits capturing a greater share of the revenue during periods of high profits;

(iii) avoids the introduction of distorting effects through the fiscal instruments;

(iv) maximizes the present value of revenue receipts by providing for appropriations during the early years of production; and

 (v) is neutral and encourages economic efficiency as a yardstick.

Table 3. A Flexible, Neutral, and Stable Fiscal Regime

What do flexibility, neutrality and stability mean?	Advantages
• A "flexible" fiscal regime is one that provides the government with an adequate share of economic rent under varying conditions of profitability.[16]	• One of the most important advantages of establishing a flexible structure (a progressive mechanism for rent extraction) is its stability over time: as market and project conditions change over time,[18] flexible fiscal systems limit the need for renegotiation.
• This type of regime targets the economic rent.[17]	• The advantage of a neutral fiscal regime is its economic efficiency. A neutral tax does not impact resource allocation. With respect to the investing company, a tax is neutral when it leaves the pre-tax ranking of possible investment outcomes equal to the post-tax ranking. With respect to a particular industry, a tax is neutral when it does not divert investments to or from that industry.
• A "neutral" fiscal regime neither encourages over investment nor deters investments that would otherwise take place.	• Stability clauses can be grouped under two categories: "freezing clauses" that maintain the contract and/or fiscal terms unchanged for the duration of the contract or for a certain period of time; and "equilibrium clauses" that allow for an adjustment of the contractual terms over time so that a change in circumstances does not damage or benefit one party to the advantage or detriment of the other.[19]
• A "stable" fiscal regime is one that does not change over a certain period of time, or one whose changes are predictable.	• In industries with long time cycles and substantial up front investments, stable and predictable contractual and fiscal terms are an important consideration in ranking investment opportunities, with obvious effects on a country's future prospects. This is particularly true for the oil and gas industry, in which long project cycles are coupled with great uncertainty with regard to resource prices and project output. • The stability of the fiscal regime also impacts business confidence and affects the level of investment in and pace of development of existing projects. • Contract and fiscal stability clauses are used in both concessionary and contractual systems. According to a recent study, of 110 countries analyzed, 77 percent offered fiscal stability protection.[20] In a recent survey conducted by Deloitte on behalf of the Aberdeen and Grampian Chamber of Commerce, contract and fiscal stability was indicated as primary factor in determining business confidence and ranking investment opportunities.[21]

---

16. There are various ways to do this. For example, one could envisage a progressive income tax and a sliding scale royalty in the case of concessionary systems; or a progressive government take linked to petroleum prices or project rate of return in the case of production-sharing arrangements. The further "downstream" a government goes to extract the rent, the less regressive the system. Signature bonuses (which are paid before a discovery is made) and royalties (which are paid whether or not a field yields a positive result) are the most regressive forms of rent extraction.

17. A fiscal regime that targets the extra revenue that is not required to persuade the investor to continue with the investment and that, if taxed away, will still allow the company to realize an acceptable return on its investment.

18. One of the most important elements of profitability of a project is the oil price level. The variability and volatility of oil prices provide for the possibility that even projects with normal profits can experience periods where excess profits are generated.

From the investing company's standpoint this means the search for a tax system that provides for:

(i) a minimum number of front-end loaded non profit-sensitive taxes;
(ii) the ability to repatriate profits to shareholders in their home countries; and
(iii) an overall policy environment that is transparent, predictable, stable and based on internationally recognized industry standards and the rule of law so that decisions can be made with reasonable confidence.

The use of flexible, neutral and stable fiscal regimes facilitates the reconciliation of these objectives. The characteristics of these regimes are summarized in Table 3.

In addition to the above described characteristics, the host government needs to take into consideration its relative position vis-à-vis other countries. In a purely competitive world, countries with favorable geologic potential, high wellhead prices, low development costs, and low political risk will tend to offer tougher fiscal terms than those with less favorable geology, low wellhead prices, high development cost, and high political risk. The economic strength and political stability of the country, oil supply balance, regional market demands, global economic conditions, and financial health of the petroleum sector also influence fiscal terms. It is commonly accepted that the level of *government take*[22] is inversely proportional to the quality and availability of investment opportunities. However, countries with harsh fiscal regimes or the greatest success probability provide no guarantees in the profitability of the project. Because the fiscal terms are only one of the elements that determine the profitability of a project, a "tough" contract may be highly profitable, while a very "favorable" contract may not be.[23]

It is important to note that good fiscal design without complementary institutional structures may still not achieve the desired goals: design needs to be within the administrative and audit capacity of the relevant institutions. Therefore, a simpler system may be more viable than a theoretically ideal but complex to manage system.[24] This is particularly important in countries that are new to the oil industry and/or have significant capacity constraints.

---

19. For a summary of the law and practice with respect to renegotiation of long-term international investment agreements in the natural resources and energy sector see Abba Kolo & Thomas W. Walde, Renegotiation and Contract adaptation in the International investment projects: Applicable Legal Principles & Industry practices, Transnational Dispute Management, Volume 1, Issue 01, February 2004.

20. As noted by Baunsgaard (2001), the data refers to contracts in existence up until 1997. Since then more and more countries have been offering fiscal stability clauses.

21. See the Aberdeen and Grampian Chamber Oil and Gas Survey, 2004–05. On the importance of fiscal stability the report concludes: "The last two years have witnessed an upturn in North Sea activity along with increased capital investment and drilling activity. Our first three surveys charted this rise in activity and the parallel rise in business confidence. A recurrent theme throughout previous surveys has been the critical importance of a stable tax regime if this rise in North Sea activity is to continue. The Government's decision to increase the supplementary North Sea oil charge from 10% to 20% jeopardizes the fiscal stability essential to sustain activity, encourage investment and maintain the North Sea's longevity. As we predicted, we can now see the first signs of the impact the Chancellor's pre-budget report has had on investment plans for the UK continental shelf."

22. The government take is the host government's share of the revenue streams associated with a particular project. It is normally expressed in percentage terms. See Chapter 6 below.

23. This is one of the reasons used to argue in favor of project specific fiscal terms as opposed to standard, sector wise, non negotiable fiscal terms (Johnston 1994).

24. Royalty and tax systems, even with sliding scale features, are used by many countries, and are usually quite easy to manage.

Fiscal Systems' Measures and Economic Indicators

To evaluate a fiscal system, governments and oil companies use different measures:

- Oil companies aim to optimize their portfolio of assets. They use economic measures to compare investment opportunities worldwide and to assess their relative risk-reward profile. During the economic life of an asset, oil companies monitor the revenue generated by it to verify that they have covered the capital investment and expenditures and that the return on capital is consistent with the risk associated with the particular asset and with the strategic objectives of the corporation.
- Host governments are interested in evaluating whether a fiscal system responds to its intended objectives. To do so, at a project level host governments use economic and system measures to assess whether the benefits—financial and social—derived from the project are consistent with its risk level and with the objectives of the government's sector policy. At a country level host governments monitor the impact of the revenue flow generated by the oil sector as a whole on the key macro-economic indicators (mainly inflation, GDP growth, balance of payments).[25]

Economic and fiscal systems measures are project-specific quantities that vary with numerous system parameters unique to the project (including, but not limited to, the size

---

25. In practice, governments in resource rich developing countries often suffer from capacity constraints that limit their ability to set up and implement rigorous monitoring systems.

and quality of discoveries,[26] the development and operational plan of the operator, the cost structure; the financing costs, discounts or premia for the particular crude oil stream[27]), as well as non-project specific variables (such as crude oil prices, inflation, currency exchange rates, local and global economic conditions, and regulatory changes). Hydrocarbon price, development cost, technological improvements, demand-supply relations, country risk, and the corporate strategy, all impact investment planning. Hence the accurate computation of the economic and fiscal system measures associated with a field largely depends on the reliability of the assumptions.[28] In effect, only at the end of a field's economic life, when all revenue, cost, royalty and tax data are known, can the profitability and the division of profits between the host government and the investors be reliably determined. In practice, due to their commercial sensitivity, cash flow and cost data are very rarely made public.

Various economic indicators are used to assess the performance of a project. The most common are the *net present value* of the project's cash flow *(NPV)*,[29] the *internal rate of return (IRR)*,[30] and the *profitability ratio (PR)*.[31] The NPV provides an evaluation of the project's net worth to the investor in absolute terms, while the IRR and the PR are relative measures used to rank projects for capital budgeting. Economic values are not intended to be interpreted on a standalone basis, but should be used in conjunction with other system measures and decision parameters. A combination of indicators is usually necessary to adequately evaluate a contract's economic performance.[32]

---

26. A detailed and realistic field description is the first and most important estimate that must be made: the size, shape, productive zones, fault blocks, drive mechanisms, etc. of the reservoir must be estimated with as much accuracy as possible since they determine the capacity of the structure and the number and location of wells. Various techniques (such as geologic conditions at reservoir level and decline curve analysis) are available to estimate production rates. However, forecast production is only used as a guideline since investment activity can dramatically alter the form of the production curve as well as recoverable reserves.

27. See R.Bacon and S.Tordo, Crude oil price differentials, October 2005.

28. It is important to underscore that the project's stage of development impacts the accuracy of the estimates and the uncertainty associated with the economic outcome of the field. On average, initial cost and production estimates may be over or under estimated by 25–50 percent compared with actual numbers; conceptual development plan estimates are normally more accurate (plus or minus 15–25 percent—Minerals Management Service, March 2004).

29. The NPV is the present value of expected future cash flow of a project. The discount rate should be a function of the riskiness of the estimated cash flows. In reality, companies often use a "hurdle rate" which represents the minimum return that the particular company is willing to accept in order for it to invest in the project. Each company has a unique risk-reward profile, hence uses a specific discount rate. The choice of what discount factor to use is an important decision for companies evaluating projects since selecting a high rate may result in "missing" good investment opportunities, while selecting a low rate may expose the firm to unprofitable or risky investments (see Allen and Seba 1993; Deluca 2003; Ehrhardt 1994).

30. The IRR measures the relative attractiveness of a project. In general terms, projects that present higher IRR should be preferred. Due to its limitations the IRR is normally used in conjunction with other profitability indices. For an in depth discussion of the IRR and of other commonly used financial measures of profitability see Brealey and Myers (1991).

31. The PR is calculated as the ratio between the NPV of the sum of project's cash flow and total capital invested in the project to the NPV of the total capital invested in the project. It measures the profitability per dollar invested and is used by companies to compare projects around the world.

32. For an extensive description of project evaluation and project financing techniques, see inter alia, Brealey and Myers (1991), Dougherty (1985), Ehrhardt (1994), Finnerty (1996), Mian (2002), and Woods (1993).

One indicator frequently referred to in sector literature is the division of profits between companies and government (the "take"). The take is a fiscal statistic as opposed to an economic measure. Because the take does not provide a direct indication of the economic performance of a field, it generally matters more to the host government than to the oil companies.

The take is often a negotiated quantity that depends upon the strength, knowledge, experience, and bargaining position of the oil company and host government, the perception of the risk associated with the field development at the time the contract was written, and the availability of opportunities worldwide.

Unlike economic measures, which are generally well-established, general confusion surrounds the application and interpretation of take.[33] In this paper, the *government take* is defined as the government's percentage of pre-tax project net cash flow adjusted to take into account any form of government participation. The government take can be calculated in discounted or undiscounted value.[34]

The take statistics for a given country offer a first frame of reference to assess whether or not the fiscal terms applicable to a contract under negotiation are in line with those that already exist in that country (Johnston 2003), or as benchmark to determine the competitiveness of a country's fiscal terms.[35] However, comparing the take of different projects and/or different countries is a very difficult and often misleading exercise because:

■ Calculating the take at project level requires: (i) ex-ante, the ability to forecast the expected cash flow for the project. As noted above, estimating the cash flow of a prospective project is highly uncertain, and even under the best conditions, is based on incomplete and often unobservable information; (ii) ex-post, the availability of information that is normally proprietary and not publicly known;

■ The same limitations apply to the calculation of the take at country level. In addition, in a given country numerous vintages of contracts are normally in force at any one

33. For more details see Johnston, Van Meurs and Seck, Smith, Wood, Allen and Seba, Barrows, and Kemp.

34. Like companies, each host government has a unique risk-reward profile/discount rate. The host government does value money in the same way as companies do. However, the host government's expected benefits should be discounted using the social discount rate. This reflects society's preferences for allocating the use of resources over time. A higher rate will attribute more weight to benefits to the current generation than to future generations. The calculation of the parameters that are necessary to determine the social discount rate involves a certain degree of value judgment. In addition, countries may have considerably different social discount rates. See Evans (2006) for a brief analysis of social discounts rates in the European Union.

35. Sector literature conventionally compares countries' fiscal systems on the basis of the government take. One use of take statistics is to calculate the possible range of the "take" for various countries using common sets of assumptions and use the result as general indicators of the relative attractiveness/efficiency of those country's fiscal systems (see for example Johnston and van Meurs). Alternatively, an attempt could me made to adjust the values that determine the take to local conditions (see the take statistics of WoodMackenzie). Before the oil shocks of 1974 and 1979, a 50 percent government take was considered a fair value but after the creation of OPEC, companies began to accept some erosion of their share of profits (Rutledge and Wright 1998). A study carried out by Petroconsultants in 1995 showed that the government take in more than 90 percent of the 110 countries examined ranged from 55 to 75 percent. Other studies have shown similar results (Johnston 1994b; Kemp 1987; Van Meurs and Seck 1995; Van Meurs and Seck 1997).

time;[36] countries typically use more than one arrangement; and contracts are often renegotiated as political and economic conditions change, or as better information becomes available.[37]

■ In industry statistics the government take is usually determined on the basis of theoretical price and cost assumptions. As noted above, the actual government take can be quite different from the theoretical average.

■ The take is inconsistent with the economic measures mentioned above, since it is frequently calculated and reported on an undiscounted basis. There can be a significant difference in the level of take depending on the manner in which the cash flow elements are discounted. For example the discounted take is normally much higher than the undiscounted one for regressive front-loaded systems.[38]

■ As the government take is made up of different elements, more or less regressive, the risk-profile, hence the attractiveness to investors, of two fiscal regimes that present the same percentage government take can be dramatically different.[39]

■ The government take does not capture the spill over effects of oil and gas projects on the economy at large.[40]

Using economic measures like the profitability index or the return on investment is also difficult as each government and each company has a unique risk-reward profile, and hence uses a specific discount rate. This of course provides the scope for negotiating contract and fiscal terms. Nevertheless, and keeping in mind the limitations expressed above, a comparison of various countries' fiscal systems on the basis of the government take, the effective royalty rate,[41] and the percentage of government participation is shown in Appendix B.

---

36. When "model" contracts are available, these are normally used as a starting point for negotiation, and the final negotiated fiscal terms are not normally disclosed or released to the public.

37. According to a study conducted by the Minerals Management Service in March 2004, each year a licensing round is launched in 25–50 countries; new model contracts or fiscal regimes are introduced in approximately 20 countries; and tax laws are revised by many countries during their annual budgetary process.

38. It is important to note that given the cash flow profile typical of oil and gas projects, an undiscounted take can be quite misleading as it would underestimate the effective government take and overestimate the effective company take.

39. The government take indicates how much of the available cash flow the government takes, but not how it takes it (D. Johnston). In addition, the take does not adequately capture the effect of, inter alia, ring-fencing provisions, reserve/lifting entitlements, and work program provisions.

40. The economic impact of industrial hyperactivity in the United Kingdom sector of the North Sea was a direct result of the "lenient" terms of the 1990s (Johnston 2001).

41. The effective royalty rate (ERR) is defined as the minimum share of revenue (or production) that the host government might expect to receive in any given accounting period from royalties and its share of profit oil. The ERR normally excludes the effects of government participation. If the contract or concession agreement has no cost recovery limit and no royalty, the host government may receive nothing in a given accounting period. This can happen even with profitable fields in the early years of production when exploration and development costs are being recovered. The world average ERR for concessionary systems is around 10 percent, whilst for PSCs it is closer to 30 percent (Johnston 2003).

CHAPTER 7

Designing Petroleum Fiscal Systems

Issues to be Considered

The host government's ultimate objective should be to design a flexible fiscal system that favors the investing companies' and the government's mutual interests by providing an equitable arrangement for both the highly profitable and the less profitable discoveries. Examples of these systems can be found all around the world: approximately 25 percent of the petro-states have some vintage of contract with R-Factor or RoR-based parameters, and the vast majority use production-based sliding scales.[42]

Decisions on the design of an appropriate fiscal framework can be supported by an understanding of how its various components influence decision making and outcomes. To this end a simplified[43] economic model of four hypothetical petroleum projects was developed to illustrate the difficulties that a country would typically face in designing a suitable fiscal framework for the development of its hydrocarbon resources. In particular, simulations were conducted to show the effect on project economics of alternative fiscal terms and their relative responsiveness to changes in economic conditions. Table 4 summarizes the key project parameters utilized in our analysis.

---

42. Statistical data provided by D. Johnston, December 2006.

43. In modeling the field economics under different fiscal systems, a number of simplifying assumptions were made. In particular: no distinction was made between intangible and tangible costs; a five year straight line amortization criteria was used for all classes of assets; investment credits (normally cost recoverable and not tax deductible,) were not considered; abandonment provisions were not included. Where the participation of a national oil company was considered, its share of expenses was carried by the contractors' group without applying any interest rate. A deterministic approach was used to calculate production levels, costs and prices. Statistical or stochastic methods could have been applied to determine the possible value distribution of the project variables, which in turn would have provided valuable information for the design of the fiscal system. Because the objective of this paper is not to optimize the fiscal system in a particular country, but merely to show how different fiscal systems respond to changes in economic and project conditions, this approach was not attempted as it would not significantly affect the result of our analysis.

Table 4. Key Project Parameters

Parameter	Field A	Field B	Field C	Field D
Recoverable Reserves	20 MBO	50.0 MBO	100 MBO	600 MBO
Peak Production Rate[44]	6.0K Bopd	15.0K Bopd	28.5K Bopd	150.7K Bopd
Field Life	20 years	20 years	20 years	20 years
Oil Price	US35/ Bbl	US$35/Bbl	US$35/Bbl	US$35/Bbl
Total Capital costs (Capex)	US$123 Million	US$234 Million	US$336 Million	US$4,615 Million
Full cycle Operating costs (Opex)	US$4.54/Bbl	US$4.24/Bbl	US$3.05/Bbl	US$2.31/Bbl

The economics of these hypothetical assets were calculated under PSC.[45] Four alternative types of sliding scales were modeled: daily production, cumulative production, R-Factor,[46] and RoR.[47] Their relative performance was assessed by allowing a selected number of fiscal and system parameters to change.[48] The results were measured in terms of break-even price,[49] NPV of the project's cash flow, IRR, PR, net present value per barrel of oil equivalent (NPV/BOE),[50] operating leverage,[51] percentage government take,[52] and saving index [SI][53]). These are summarized in Table 5. Detailed calculations are shown in Appendix D, Tables 22 to 25.

---

44. When simulating the impact of variations in production levels, the same percentage was applied through out the production horizon (that is, no adjustments were made to the production rate to take into account facilities specifications and/or reservoir management needs).

45. Theoretically it is possible to exactly replicate a particular fiscal regime using different combinations of fiscal instruments—for example a production sharing contract can be replicated by a combination of royalties and taxes. Hence the choice between contractual systems and concessionary system mainly depends on the country's administrative capacity or on the objectives of its sector policy (Baunsgaard 2001). For this reason, concession agreements were not modeled in this paper as this would not significantly affect the analysis.

46. The R-Factor was calculated as the ratio between after-tax revenues and total project costs (capital expenditure and operating costs). As it is the case for many other system parameters, the definition of R-Factor tends to be country (sometimes contract) specific. Therefore, one should be cautious in comparing fiscal parameters among countries/contracts as their effect on project economics can be quite different.

47. See Appendix B, Table 11.

48. To simplify the interpretation of the results, only one parameter at a time was allowed to change. In reality, there are dependency relationships among parameters. The likelihood, magnitude, and timing of changes in technical and economic parameters have different effects on project economics, and on the overall performance of the system. A stress test was, however, carried out for all fiscal models by calculating the project's NPV at different discount rates resulting from decreasing the production level and price by 20 percent and increasing capex and opex by 20 percent.

49. The minimum level of gas price that causes the project's NPV to become zero.

50. This indicator allows companies to compare investments around the world, irrespectively of the size of the project.

51. The operating leverage was calculated as the ratio of the net present value of total cost to the net present value of gross revenue. Both flows were discounted at 10 percent. The higher the operating leverage, the more exposed the project profitability is likely to be to a fall in prices.

52. The government take was calculated on an undiscounted and on a discounted basis. To simplify the comparison with the contractor's take, all cash flows were discounted at 10 percent. In reality, the government's cash flow should be discounted at the social rate (see note 34 above). This is likely to be lower than 10 percent, thus increasing the percentage government take.

53. The Saving Index (SI) is defined as the part of an additional one dollar in profit (arising from a one dollar saving in cost) that accrues to the contractor. It measures the degree to which the contractor will benefit from a reduction in costs (see Johnston 2003).

Table 5. Fiscal System Indices

	Field A			
	Fiscal Model 1	Fiscal Model 2	Fiscal Model 3	Fiscal Model 4
Contractor's Cash Flow (NPV10%)	70.6	70.6	73.5	79.9
Break-Even Price	18.64	18.64	17.48	17.07
Project's IRR	26.0%	26.0%	28.2%	29.0%
NPV(10%)/BOE	3.53	3.53	3.67	3.99
PR(10%)	0.60	0.60	0.62	0.68
Operating Leverage (%)	43.2%	43.2%	43.2%	43.2%
Government Take (%)	57.6%	57.6%	55.9%	52.0%
Saving Index (US$)	0.53	0.53	0.52	0.56

	Field B			
	Fiscal Model 1	Fiscal Model 2	Fiscal Model 3	Fiscal Model 4
Contractor's Cash Flow (NPV10%)	213.7	213.7	212.4	211.0
Break-Even Price	15.05	15.05	14.13	13.79
Project's IRR	33.4%	33.4%	36.6%	36.4%
NPV(10%)/BOE	4.27	4.27	4.24	4.22
PR(10%)	0.88	0.88	0.88	0.87
Operating Leverage (%)	35.5%	35.5%	35.5%	35.5%
Government Take (%)	54.6%	54.6	54.9%	55.2%
Saving Index (US$)	0.53	0.52	0.49	0.49

	Field C			
	Fiscal Model 1	Fiscal Model 2	Fiscal Model 3	Fiscal Model 4
Contractor's Cash Flow (NPV10%)	515.3	469.7	456.0	363.4
Break-Even Price	10.56	11.32	10.00	9.62
Project's IRR	49.7%	49.4%	55.2%	50.4%
NPV(10%)/BOE	5.15	4.70	4.56	3.63
PR(10%)	1.50	1.37	1.33	1.06
Operating Leverage (%)	24.9%	24.9%	24.9%	24.9%
Government Take (%)	53.0%	57.1%	58.4%	66.8%
Saving Index (US$)	0.52	0.47	0.42	0.36

	Field D			
	Fiscal Model 1	Fiscal Model 2	Fiscal Model 3	Fiscal Model 4
Contractor's Cash Flow (NPV10%)	386.4	(14.4)	1,751.8	2,423.0
Break-Even Price	29.41	35.23	20.30	19.44
Project's IRR	12.1%	9.9%	19.2%	21.2%
NPV(10%)/BOE	0.64	(0.02)	2.92	4.04
PR(10%)	0.10	(0.00)	0.45	0.62
Operating Leverage (%)	47.4%	47.4%	47.4%	47.4%
Government Take (%)	91.6%	100.3%	61.7%	47.1%
Saving Index (US$)	0.34	0.18	0.51	0.64

Our simplified analysis illustrates that the anticipated size and distribution of production in a given geological province is a key element in the design of a fiscal system. This can be seen by applying the same fiscal model to different size fields and comparing its performance (Appendix D, Tables 22 to 25). Furthermore, for all fiscal models/fields analyzed in this paper, variations in production from the base case level considerably impacted project economics (plus or minus 30 percent for the smaller size fields, plus or minus 23 percent for the medium size field, and plus or minus 43 percent for the large size field[54]). Similar results were obtained for price variations. Decreases in production and prices resulted in large percentage variations in project NPV because of the rigidity of capital investment. The higher the project's operating leverage, the larger the impact of a variation in price or production level. In our models a variation in the level of production had the lowest effect on the project's NPV for Field C (24.9 percent operating leverage), while Field D (47.8 percent operating leverage) was affected the most. These are very important considerations in the design of a fiscal system, as market prices and geological conditions can be estimated only with a high degree of uncertainty. Therefore, companies undertaking capital intensive and complex projects (for example deep water or gas projects), or those in frontier or remote areas, or risk-adverse or smaller companies would logically prefer fiscal systems that provide a cushion in case of adverse conditions.[55] Projects with high operating leverages, all other parameters being equal, are relatively more exposed to the risk of losses under regressive fiscal regimes (Kretzschmar and Moles 2006). When project financing is involved, a fiscal regime that is less sensitive to variations in project economics will increase the perception of risk and, ultimately, the average cost of capital and the exploration and development thresholds.

Because capital expenditure occurs mainly in the initial phase of a project, variations in its level have a large impact on project economics, especially when a cost recovery limit is imposed and/or the state participating interest is on concessional terms.[56] In general terms, higher cost recovery limits allow the contractor to achieve payback of its investment faster. However, when sliding scales are used to determine the percentage of profit oil split (or the tax rate), in some cases higher cost recovery limits may lower the contractor's full cycle discounted cash flow. This would depend on several factors, including the level of saturation of the system, the operating leverage, the discount factor, and the steepness of the sliding scale vis-à-vis the changes in the project IRR. In Appendix E, Graphs 1 to 4 show the effect on project profitability of different levels of cost recovery limit for the fiscal systems modeled in this paper.

In designing fiscal systems, it is important to create an alignment between the contractors' and host government's interests. In this context, creating incentives for cost sav-

---

54. Because of the misalignment between the project's daily and cumulative production levels and the production thresholds used to calculate the royalty and the profit oil split, the variation in the project's NPV was much bigger for the large size field under Fiscal Model 1 and 2.

55. The expected average field size and likely development solutions provide valuable information for the design of fiscal system and/or fiscal incentives even with respect to mature areas.

56. It is worth noting that during primary recovery, only a small percentage of the initial hydrocarbons in place are produced, typically around ten percent for oil reservoirs. Secondary and tertiary recovery methods that may be employed once the reservoir primary recovery—natural drive or artificial lift—reaches its limits are quite expensive. Therefore, to avoid leaving technically producible reserves in the ground, the fiscal system should not discourage this type of investment.

ings is an important objective. Typically, the contractor would have an incentive to save (especially during the exploration phase). The extent of the benefit depends on the profit-based elements of the fiscal system (for example, profit sharing and taxes). It also depends on the timing of the saving, as the present value of a dollar saved today is higher than that of a dollar saved tomorrow.[57] Table 6 below summarizes the effect of a 20 percent variation in capital expenditure for the fiscal models modeled in this paper. In general terms, fiscal systems that have a low contractor's marginal take are more likely to create a lower incentive to saving because the majority of the savings will be transferred to the government. In extreme but rare cases, inefficiencies in the fiscal system may encourage the investor to spend more than it otherwise would.[58] To mitigate this type of inefficiency, the host government should ensure that thresholds and triggers are not too wide, that is, changes in thresholds corresponds to changes in the project IRR and changes in triggers do not discourage savings by capturing the whole of the project upside. This is particularly important for complex projects with high capital investment and long implementation periods.

Table 6. Contractor's and Host Government's NPV Variation								
	Field A		Field B		Field C		Field D	
	Contr.	Govt.	Contr.	Govt.	Contr.	Govt.	Contr.	Govt.
20% increases in Capex								
- Fiscal Model 1	−65%	−35%	−65%	−35%	−64%	−36%	−57%	−43%
- Fiscal Model 2	−65%	−35%	−65%	−35%	−64%	−36%	−52%	−48%
- Fiscal Model 3	−50%	−50%	−61%	−39%	−50%	−50%	−50%	−50%
- Fiscal Model 4	−22%	−78%	−17%	−83%	150%	−250%	−63%	−37%
20% reduction in Capex								
- Fiscal Model 1	65%	35%	65%	35%	64%	36%	58%	42%
- Fiscal Model 2	65%	35%	65%	35%	64%	36%	49%	51%
- Fiscal Model 3	44%	56%	11%	89%	−18%	118%	54%	46%
- Fiscal Model 4	34%	66%	−26%	126%	−104%	204%	35%	65%

The choice of trigger rates and thresholds is a key issue for all fiscal models. It is quite unlikely that a particular set of triggers or thresholds would be able to optimize the government take under all possible scenarios. For example, if the thresholds for triggering higher profit oil/gas splits are too wide, the system may not efficiently capture the economic upside of a project.[59] This can be seen in Appendix F, which shows the effect on government take and project IRR of applying different daily production thresholds to calculate the royalty and the profit oil split for Field A.

---

57. For this reason the contractor's actual saving benefit in present value terms is quite likely to differ from the SI.

58. See Johnston 2003. This is gold plating in its "purest" form. Fiscal Model 4 (and to a lower extent Fiscal Model 3) applied to Field C exemplifies the concept.

59. Similar consideration applies to sliding scale royalties and taxes.

The comparison of NPV and government take shown in Appendix D illustrates the complexities of defining efficient profit oil splits between the host government and the investor. In particular:

▇ The daily and cumulative production thresholds necessary to trigger higher profit oil splits in favor of the government were never reached when Fiscal Models 1 and 2 were applied to Fields A and B, and the percentage profit oil split remained the same for the economic life of the two fields (see Tables 22 and 23).

▇ On the other hand, when applied to Field C, Fiscal Model 2 was able to capture more revenue for the host government than Fiscal Model 1. This is because under Fiscal Model 2, the field reached its second profit sharing and royalty threshold at approximately 50 percent of total production, while under Fiscal Model 1, the daily production level was only slightly above the first threshold for a short period of time (see Table 23).

▇ Because of the high level of capital investment, the application of Fiscal Model 2 to Field D made the project uneconomic (see Table 24).

▇ There were no significant differences between R-Factor and RoR-based profit split in the first four years of production and between the seventh and the eleventh year of production for Field A.[60] This is because during these periods, changes in threshold in both the R-Factor and the RoR-based models corresponded to changes in the project's internal rate of return.

These examples illustrate that in order to capture a suitable share of profit oil the host government would need to make reasonable assumptions about the size and profile of a typical project, as well as to determine the typical variability in key project parameters. This would allow it to determine a representative distribution of R-factors, RoRs, or other parameters chosen as thresholds and triggers, and to set appropriate floors and ceilings for such thresholds and triggers. Unfortunately, historical data, even if determined with reference to specific geological basins and project locations,[61] often do not provide sufficient guidelines.[62] Therefore, a certain degree of "art" in defining the thresholds and triggers would still be required. Anticipating project profitability for a specific asset is a difficult exercise. Imagine how difficult it would be for a host government to attempt to define profit oil splits that can apply universally to all projects in the country. This is one of the reasons why the profit oil split is normally the subject of negotiation.

The majority of existing production sharing contracts (and concession agreements) uses sliding scales based on cumulative or daily production levels. In some cases, different thresholds and trigger rates apply depending on, for example, the water or well depth. In

---

60. Similar results were observed with respect to Field B and Field C, where the two fiscal systems behaved in exactly the same manner except, respectively, between the fifth and the seventh years and the third and fifth years of production. On the other hand, the two Fiscal Models behaved quite differently when applied to Field D.

61. In addition, assumptions about the level and volatility of oil price—which are among the most important elements of project profitability—would still to be made.

62. Establishing the fiscal terms for service contracts, concessionary agreements and PSCs for non-exploration assets is less problematic as the risk profile of the asset, and often its likely costs, are sufficiently known.

some production sharing contracts the production based profit oil/gas split is further linked to the level of oil/gas prices and/or the R-Factor. Sliding scale terms introduce flexibility in fiscal systems. This theoretically allows small and large fields to be developed on equitable terms. In reality, as shown in our simplified models, the neutrality of the system largely depends on how the thresholds are defined, and how closely they relate to the profitability of the underlying project.

Mathematically, it is always possible to design thresholds and triggers of a sliding scale based on production levels that match the changes in project economics. Because this can be done only at the end of the life of any given project and is bound to be different for each project, the use of RoR and R-Factor triggers is likely to be more efficient at sharing the project's upsides and downsides between the contractor and the host government. Furthermore, because of their flexibility, R-factor and the RoR-based models generally have a lower break-even price (see Table 7 below), which makes them more attractive to the contractors and less risky candidates for project financing.

Table 7. Break-Even Price				
	Fiscal Model 1	Fiscal Model 2	Fiscal Model 3	Fiscal Model 4
Field A	18.64	18.64	17.48	17.07
Field B	15.05	15.05	14.13	13.79
Field C	10.56	11.32	10.00	9.62
Field D	29.41	35.23	20.30	19.44

In Appendix, Graphs 5 to 8 show the sensitivity of government take and project profitability to changes in prices for the fiscal systems modeled in this paper. In general terms, production based sliding scales are more regressive (less neutral to investment decisions) than are R-factor and RoR-based sliding scales, as the split remains the same even if important changes in project economics should occur. On the other hand, these systems are easier to administer and may prove reasonably efficient in sharing the rent between the contractor and the government when project uncertainty is low, especially if used in conjunction with price indices.

The impact on project economics of the government's participation through the NOC deserves special consideration. If concessional conditions apply to the government back-in interest (if the government does not pay its way in, or pays it only partially) this would have implications for the contractor's NPV.[63] Furthermore, because the contractor is allowed to recover expenses (its share and the carried) with a limited or unlimited carry forward,[64] this may result in an implied borrowing rate for the host government that is higher than

---

63. Normally the State/NOC is carried through the exploration phase (rarely through development). Exploration costs may or may not be reimbursed to the contractor. In some cases interest may apply to unrecovered costs. The State/NOC share of participation is normally paid out of production.

64. Usually unrecovered expenses are carried forward to the next fiscal year until they are fully recovered (unlimited carry forward) or until they are allowed for recovery (limited carry forward). The carry forward affects both the calculation of profit oil split and of the corporate tax, although the mechanism may be different.

its marginal borrowing rate. In addition, unrecovered expenses affect the calculation of R-Factor and RoR, which in turn may affect the level of government revenue when profit oil split/taxes are determined on these bases. In these cases, the NOC carried participation may mitigate the effect on a contractor's NPV of a cost overrun (or improve the benefit to the contractors in case of cost savings). This effect and its magnitude would, inter alia, depend on the relationship between project IRR and discount rate, and it would not occur when the project IRR is lower than the discount rate. Therefore, when a carried interest is involved, the decision to exercise the back-in option, and the consequent use of public resources, needs to be evaluated in light of the overall macro-economic objectives and resource allocation priorities of the government. In Appendix G, Table 26 shows the effect of a 30 percent[65] participation of the NOC carried through exploration.[66]

Depending on their overall fiscal policy needs, host governments seek different levels of front-loading. To achieve their objectives while maintaining a reasonable level of investment incentives, it may be necessary to accept a tradeoff between regressive features (royalties, cost recovery limits) and progressive features (RoR, R-Factor based taxes or production sharing). Although progressive regimes are most successful in optimizing the government take under varying economic conditions, they may increase revenue volatility. Various risk management strategies exist to smooth revenue volatility, the costs and benefits of which need to be carefully considered.[67]

---

65. Often the contract provides for the host government to have the option to participate, directly or through its NOC, up to a certain percentage participating interest. The percentage participation of the State/NOC affects the project IRR for the contractor. It is important to note that the IRR is one of the parameters used by oil companies to rank their investment opportunities. Companies set a target rate(s) that reflects the project risk and the investor's corporate profile. All other things being equal, investment opportunities with an IRR below the target rate are not likely to be considered. Although target rates are unique to each company, a fifteen percent target rate would not be uncommon. This should be taken into consideration by the host government in determining its level of participation in any given asset.

66. In all our models, no interest rate was applied to unrecovered expenditure related to the NOC carry. The NOC and the contractor were assumed to pay the same corporate tax rate. In reality, often profit oil splits are calculated at project level, while corporate taxes are calculated at company level (each co-venturer pays taxes separately). Furthermore, the existence and extent of ring fencing affects the overall level of tax receipts.

67. These range from hedging to the use of special reserve funds (SRF). The analysis of the cost and benefits of alternative risk management strategies is beyond the scope of this paper. In general terms, policy makers will be reluctant to take the political risks of a hedging program. Therefore the creation of an SRF may be preferred, especially if the government is able to implement a stable long-range fiscal plan designed to create and preserve a SRF's balance at such a level that it can be used to effectively insure the state against revenue fluctuations. If, given the expected level of revenue and the spending needs of the country, a SRF fund may not be able to provide such insurance, a hedging program may need to be considered.

CHAPTER 8

Conclusion

Countries compete with each other to attract foreign investment to develop their natural resources. To achieve this objective, they must assess their position in the global marketplace and evaluate their particular situation, boundary conditions, concerns and objectives. In evaluating options to encourage oil exploration and production activities, host governments should focus on measures that: (i) materially improve the economics and/or reduce the investment risk, (ii) involve low compliance and administration costs; (iii) address market deficiencies; (iv) minimize distortionary effects; and (v) are consistent with the country's macro-fiscal policy and with local development objectives.

Although not all countries have made the same legal and regulatory choices, nearly all have established sector specific legislation and regulation in line with their constitution and with the rest of the country's body of laws. One advantage of this approach is its transparency and its objectivity: by establishing the boundary conditions for the award of petroleum rights and defining the authority and procedures for such award, system inefficiencies and the scope for discretional behavior are greatly reduced. Whether contractual or concessionary systems are used, key elements considered by potential investors in comparing investment opportunities include clarity and simplicity of terms; objectivity of rules and their enforcement; and neutrality, equity, efficiency, and stability of fiscal terms.

The fiscal regime, especially when complemented by broader sector reforms, has been used by many countries to convert government policy into economic signals to the market and influence investment decisions. Depending on its overall fiscal policy needs, the host government may seek different levels of front-loading at different points in time. In order to achieve its objectives while maintaining a reasonable level of investment incentives, the government would need to seek a tradeoff between regressive features (royalties, cost recovery limits, exploration tax) and progressive features (RoR, R-Factor-based taxes, or production sharing).

One of the key challenges of fiscal policy is develop a system that is able to allocate risks equitably. To meet this challenge, policy makers need to take into account the divergent interests of companies and governments. As risks can differ substantially for different projects and countries and over time, a fiscal regime that provides optimal outcomes under all circumstances is not likely to be developed. Although this may justify a case by case approach, this would hardly be efficient given the usually large number of projects and the often limited administrative capacity of the host government. It is therefore desirable to build enough flexibility into a system to allow for automatic adjustments to unforeseen changes and to minimize the need and cost of negotiations and/or renegotiations. Intuitively, this would imply the optimization of fiscal revenue at the country level as opposed to the optimization of fiscal revenue at the project level.

Many petroleum fiscal systems around the world exhibit some form of flexibility. Very few of them effectively and efficiently target the economic rent, i.e. are neutral to investment decisions. Fiscal systems that use sliding scales based on daily or cumulative production targets are insensitive to changes in prices and costs. So in a dynamic environment like the one that characterizes the oil industry, these systems are more likely to produce a misalignment of interests between the host government and the investors, as the recent surge in contracts renegotiations suggests. On the other hand, these systems are relatively easy to administer and may prove reasonably efficient in sharing the rent between the contractor and the government when project uncertainty is low, especially if used in conjunction with price indices.

R-Factor and RoR-based fiscal systems lower the project specific risk by introducing flexibility in the fiscal package to suit the profitability of the particular project. Because of their flexibility, these types of arrangement are more likely to encourage the development of marginal fields, or of complex projects with a long lead time for implementation. In addition, the use of R-factor and RoR-based systems normally lowers the break-even price of a project. This in turn makes these projects more attractive to the contractors and less risky as candidates for project financing.

The choice of trigger rates and thresholds is a key issue for all fiscal systems. It is quite unlikely that a particular set of triggers or thresholds would be able to optimize the government take under all possible scenarios. In order to define relevant thresholds and triggers, the host government would need to make reasonable assumptions about the size and profile of a typical project, as well as to determine the typical variability in key project parameters. This would allow it to determine a representative distribution of R-factors, or RoRs, or other parameters chosen as thresholds and triggers, and to set appropriate floors and ceilings for such thresholds and triggers. The efficiency and neutrality of the fiscal system largely depends on how closely triggers and thresholds relate to the profitability of the underlying projects. In general terms, wide thresholds may not efficiently capture the project rent, and steep trigger rates may have distortive effects on investment decisions.

The issue of government participation (the back-in option) in oil and gas exploration and production activities deserves special consideration. Nearly half of the countries around the world allow some form of participation through the NOC, the oil minister, or other government entity. Countries that use PSCs are more likely to use government participation as means of rent extraction. Governments that allow participation may or may not reimburse exploration costs to the contractor. Those who do not, normally allow the contractor to recover expenses (its share and the "carried") with a limited or unlimited carry forward.

From a purely financial standpoint, this has implications for the contractor's NPV and IRR. In some cases, the carry may result in an implied borrowing rate for the government that is higher than its marginal borrowing rate. Un-recovered expenses affect the calculation of R-Factor and RoR, which in turn may affect the level of government revenue when profit oil split/taxes are determined on these bases. Therefore, when a carried interest is involved, the decision to exercise the back-in option, and the consequent use of public resources, needs to be evaluated in light of the overall macroeconomic objectives and resource allocation priorities of the government.

Even when a flexible petroleum fiscal regime is established, the host government would still need to regularly assess its performance and to adjust the relevant parameters as needed so that the fiscal regime applicable to future projects reflects changes in market conditions, government policy, and geological and country risks. Finally host governments would need to periodically re-assess the impact of their petroleum fiscal system on the overall macroeconomic framework to ensure it encourages the efficient and effective use of resources.

APPENDIXES

Key Elements of Successful Petroleum Legal Frameworks

Area	Key components
Government authority	Ownership of natural resources; powers granted to government officers; enforcement; penalties and fines; the authority to negotiate contracts; the taxing authority, and approvals authorities.
Access to the acreage	Qualifications for authorization to explore, develop, produce and process; areas closed to mineral activities; areas subject to special controls or conditions; right of ingress and egress; resolution of conflicting land disputes; and the relation between surface and subsurface right holders.
Exploration and production rights and obligations	Extent of the exploration and production area; duration of the term for exploration and production rights; renewal of exploration and production rights; unitization; cancellation or termination of a right; area relinquishment; minimum work programs; security of tenure; reporting; transferability of rights and mortgageability; surface fees.
Protection of the environment	Environmental impact assessment; environmental impact mitigation; social or community impact; monitoring and reporting; abandonment liability; reclamation; and environment sureties.
Fiscal Terms	State participation; royalties; production sharing rate and base; custom duties; income tax rate and base; special petroleum taxes; other levies and taxes; gas production incentives and other incentives; ring fencing; and stability clauses.

Tax and Non-Tax Components of Petroleum Fiscal Systems

Table 8. Royalties

How do they work?

- Royalties have historically been the most common method used by governments to gain revenue from the exploitation of the nation's mineral endowment.

- Royalties are based on either the volume (*"unit"* or *"specific"* royalty) or the value (*"ad valorem"* royalty) of production or export.[68]

- Unit royalties impose burdens that vary in inverse relation to changes in market price, while ad valorem royalties vary in direct relation to price for any given level of production or sale.

- In the petroleum industry, royalties are typically calculated on a net-back basis:. the price base for royalty calculation is adjusted from the point of export to the wellhead by deducting transportation and other marketing costs.

Advantages and Disadvantages to Host Governments

- Royalties are attractive to governments because they ensure an upfront revenue stream as soon as production starts.

- As they are attached to production or sales, they can be estimated with a reasonable degree of predictability.

- They are comparatively easy to calculate, collect, and monitor.

- Royalties are a regressive form of taxation.[69] High levels of royalties distort investment decisions and may encourage uneconomic choices. To mitigate their regressiveness, some countries apply sliding scale royalties based on production levels or sales values, water depth or well depths, or R-factors.

(continued)

Table 8. Royalties (*Continued*)

Effect on Investment Decisions

- Royalties have a tendency to distort the levels of recovery, although this effect is relevant only when the royalty is the most important part of the tax rent and when important differences in quality occur in crude oil or gas produced from a given contract area. In particular:

  - Unit royalties reduce the effective price by the same nominal amount each year. Since the net present value of the royalty decreases over time, investors will have an incentive to prefer future production over current production when future prices are expected to increase. In addition, a royalty imposed on the volume of production or sales may encourage the investor to delay the production or sale (subject to technical considerations) of the lighter, sweeter crudes or higher heating content gas if the discounted value of future prices is expected to increase.

  - Ad valorem royalties reduce the discounted price of crude oil or gas by the same percentage in each year. Therefore, if the prices are expected to rise in real terms, investors would prefer increasing production (subject to technical considerations) in the present.

- As royalties are payable whether or not the project is profitable, they can constitute a major deterrent to investment.

- By increasing the economic cut-off rate, royalties reduce the economic life of a project.[70]

---

[68]Some countries link the royalty rate to parameters like average daily production, oil price, water depth, field location, depth of reservoir, and crude oil quality.

[69]For a detailed description of royalty practice in various countries, see Otto (1995).

[70]The impact of this is to leave in the ground hydrocarbons that would otherwise be produced.

Table 9. Ring Fencing

How do they work?

- *Ring-fencing* is an industry-specific feature. This refers to the delineation of taxable entities. While corporate income tax normally applies at company level, in the petroleum sector the taxable entity is often the contract area or the individual project. When ring fencing applies at contract area or project level, income derived from one area/one project cannot be offset against losses from another area/project. Another type of ring fencing separates upstream from downstream operations. Usually all costs associated with a given block or license must be recovered from revenue generated within that block: the block is ring fenced. However, some countries allow exploration costs to cross the fence.[71]

Advantages and Disadvantages to Host Governments

- The objective of ring fencing is to protect the level of current tax revenue and, to some extent, level the playing field by treating newcomers and existing investors equally. The disadvantage of ring fencing is that it does not incentivize exploration and investment activities. However, by allowing costs to cross the fence, the host government may end up subsidizing unsuccessful exploration.

- Some countries allow the consolidation of upstream, transportation and processing activities, and an array of arrangements is used for LNG projects, which involve various degrees of consolidation. Other countries have preferred to maintain the integrity of their tax systems and to provide similar level of incentives through the definition of transfer prices or through other form of incentive.

Effect on Investment Decisions

- The relaxation of ring fencing can provide a strong financial incentive to investors, especially those who have existing production or are in a tax paying position. The existence of a cost recovery limit may enhance the importance of this type of incentive.

---

[71]In New Zealand, a 100 percent deduction is given for exploration expenditure in the year in which it is incurred; development expenditure is allowed as a deduction over 7 years from the date of expenditure for offshore wells, and any losses arising are not ring fenced either to permits, fields, or even the trade. That is, losses can be offset against any New Zealand income of the company or group of companies.

Table 10. Corporate Income Tax

How do they work?

- Some countries include the petroleum industry within the standard corporate income tax regime, although they may use a higher tax rate to capture more rent.[72] Under this method, taxes are only due when annual revenue exceeds some measure of costs and allowances. Therefore, the key elements of this tax form are the definition of taxable income and the rate applied to it.[73] In their traditional formulation (a fixed tax rate) corporate taxes are relatively regressive, as their burden in percentage terms remains the same at different levels of profitability.

- To ensure that the host government shares the upside if a project becomes very profitable, more and more countries have adopted progressive income tax rates. This is done by using stepped tax rates linked to parameters like the crude oil price, the volume of production, the sales value, and so on. These are "add-on" to conventional proportional income tax.

- In some countries the investor's income tax is paid by the government out of its share of production.

Advantages and Disadvantages to Host Governments

- Because corporate income taxes are well defined in the country's tax code, their assessment, collection, and monitoring can be more easily accommodated within the country's existing systems, thus lowering the government's administrative burden.

- Progressive income taxes tie the level of taxation to parameters that are linked to the level or activity or the price of crude oil or gas. This allows the host government to partake in the project's upsides when economic conditions are more favorable.

Effect on Investment Decisions

- The parameters normally used to determine the progressive rates of income tax are not necessarily fully correlated with the investors' return on investment. Hence this type of corporate tax may not be neutral for investment decisions.

- In countries where the tax is paid by the government (or the NOC) on behalf of the contractor, consideration should be given to structuring the tax so that they can be treated as if paid directly by the contractor for home country tax credit purposes. As the contractor is not affected by changes in tax rates, these types of agreements are generally quite stable.

---

[72]As petroleum companies operate on a multi- and/or cross-national level, it is important that the host government introduce safeguards to avoid profits being transferred to jurisdictions with lower corporate tax rate (for example, by imposing restrictions on deductible interests or inter-company charges, or requiring arms-lengths transactions).

[73]In defining the applicable rate, government should be mindful that home nation treatment of foreign earnings is ultimately of importance to investors. Rates that are too low merely transfer tax revenue to the treasury of the investors' home countries. Therefore, if incentives are to be provided, adjustments in the definition of taxable income may prove more effective. See Thomas A. Gresik, April 2001.

Table 11. Resource Rent Tax

How do they work?

- Resource rent taxes tie taxation more directly to the project's profitability. In its pure form, taxes are deferred until all expenditures have been recovered and the project has yielded a predefined target return. Then a very high marginal tax is applied to all subsequent operating revenue. Basically, the project is granted a tax holiday compared with conventional tax regimes in anticipation of exceptionally high governmental returns over time. There are two main systems:
    - R-Factor based systems, which are linked to the payback of an investment (the ratio of cumulative after tax[74] receipts to cumulative expenditures—capital expenditure and operating costs),[75] and
    - Rate of Return (RoR) based systems, which are linked to the project's return on investment (hence they take into consideration the time value of money and apply when a target internal rate of return has been achieved[76]).

  Some countries have adopted a stepped resource rent tax schedule with incremental brackets to smooth the shift to the higher tax rates.
- In some countries, the investors' income tax is paid by the government out of its share of production.

Advantages and Disadvantages to Host Governments

- The main advantage of a resource rent tax is its neutrality[77] (at least in theory[78]). The disadvantage is that it only provides income to the government when the target payback or rate of return is reached. This can be avoided by combining the resource rent tax with a royalty and/or a normal corporate income tax. The key issue then becomes that of defining an efficient target rate. This is a complex issue as it depends on the specific characteristics of the project, as well as exogenous conditions. Resource rent taxes are comparatively more difficult to assess and monitor. Therefore, the administrative cost of maintaining this system largely depends on the capacity of the host government.

Effect on Investment Decisions

- Resource rent taxes are relatively neutral to investment decisions. This depends on how close the target rate is to the investor's discount rate, which in turn reflects the project risk and the investor's corporate profile.

---

[74]In some countries pre-tax values are used to calculate the R-Factors. Examples can be found in Colombia, Malaysia (R/C index), and India (Investment Multiple post NELP V contracts).
[75]The R-Factor is calculated in each accounting period. Once a threshold is crossed, the new tax rate will apply to the next accounting period. The cash flows used in determining the R-Factor do not take into consideration the time value of money (they are undiscounted). In some royalty and tax arrangements, a stepped corporate income tax rate is determined on the basis of a range of R-Factor values (as, for example, in Chad). In some cases the R-Factor is used to determine the royalty rate (as, for example, in Tunisia). In some Production Sharing Contracts, the R-Factor is used to trigger the percentage of the government's share of profit oil (as, for example, in Qatar).
[76]In RoR-based systems net annual cash flows are compounded at the target RoR rate and carried forward until the cumulative amount becomes positive. When the investor has recovered the initial investment plus the target rate, the tax kicks in. Theoretically the target RoR rate should represent the minimum rate to encourage investment.
[77]The neutrality of a tax can be assessed by its impact on the resource allocation. With respect to the investing company, a tax is neutral when it leaves the pre-tax ranking of possible investment outcomes equal to the post-tax ranking. With respect to a particular industry, a tax is neutral when it does not divert investments to or from that industry.
[78]This depends on how close the target rate is to the investor's discount rate. The Brown tax levied on the investor's cash flow is the most neutral form of taxation, as it provides for subsidies when the project cash flow is negative and captures the excess rent when the cash flow is positive. In practice, this tax has never been applied in its purest form, although some countries do provide subsidies to corporations, and these have an effect similar to the Brown tax. For a discussion of the economic efficiency of resource taxation policy, see Garnaut and Clunies Ross (1975).

Table 12. Import and Export Duties

How do they work?

- Import duties apply to all material and equipment imported in a country. In the past, these were used to provide protection for locally produced goods.

- Almost all countries have some sort of trade duty system, but in the oil industry import duties have had a limited use as fiscal tools (local content provisions have largely substituted the use of import duties to protect local industries).

- The majority of countries provide exemptions from import duties on material and equipment destined to oil and gas operations. In some cases, the exemption is granted throughout the duration of the relevant Production Sharing Contract or Concession Agreement; in others, it is limited to the exploration and development phase.[79] Some countries provide a blanket exemption; others limit the exemption to a specific list of materials and equipment. Exemptions for temporary import of equipment are the general practice in all producing countries.

- Because export duties distort the price of export and domestic supplies, they are normally not levied on oil and gas.[80]

Advantages and Disadvantages to Host Governments

- For host governments, import duties provide a source of revenue from the very beginning of project operations. On the other hand, because of the nature of administering such duties, lower level government officers often have to classify the goods. This may create delays in processing the goods and may increase the potential for rent-seeking behavior.

- The use of lists of exempted material and equipment often increases the customs authority's administrative burden.

- Because equipment and material originally imported for use in one project area may be used in other project areas, the grant of exemptions based on the destination to a particular project area often generates inefficiencies.

Effect on Investment Decisions

- Given the very substantial import needs during the exploration and development phase of a project, the payment of import duties on material and equipment has a direct impact on project economics by reducing the net present value of the project and increasing its risk profile. For this reason, the existence of custom duties exemptions, at least in the early stages of a project, is of great value to investors.

[79]However, the administration of a mixed system can prove very difficult and expensive.
[80]Russia is among the notable exceptions.

Table 13. Value Added Tax

How do they work?

- Value Added Tax is normally levied on a destination basis, i.e. imports are taxed and exports are zero rated. For this reason, oil and gas projects would normally be in a tax credit position[81].

- The majority of producing countries exempt or negate the effect of Value Added Tax on projects that export. This is done by providing some sort of credit, refund, exemption, drawbacks or deferrals at least during the initial phases of a project and/or to at least some type of purchases.

Advantages and Disadvantages to Host Governments

- Depending on the choice of system (whether outright exemption or some form of refund, credit, drawback or deferral) the administration of Value Added Tax for oil and gas projects can be quite complex. In particular, if the capacity is not in place to administer a refund-based system, a sector specific exemption or an exemption limited to certain specialized inputs used exclusively in the oil and gas industry may be more efficient.

(*continued*)

Table 13. Value Added Tax (*Continued*)

Effect on Investment Decisions

- Value Added Tax has approximately the same effect on the investor's cash flow/return on investment as import duties. For this reasons, fiscal systems that provide exemptions in respect of at least specialized inputs are preferred by investors.

---

[81]This occurs in cases where production (that is zero-rated for VAT purposes) is only marginally sold in the local market. In these cases, the oil company would find itself in a continuous credit position, claiming refunds on VAT paid on purchase of goods and services.

Table 14. Surface Taxes

How do they work?

- Surface fees are generally paid annually on the basis of the size of the property under lease. Different fees normally apply for exploration and production acreage. Surface fees are set at a nominal amount. Their aim is to discourage investors from holding on to acreage without exploring it.

Advantages and Disadvantages to Host Governments

- Surface fees are easy to calculate, collect and monitor. They have the advantage of providing a source of revenue, albeit limited, during each phase of a project life. When a government upstream agency exists, surface fees are often collected by the agency that uses the revenue flow to cover its administrative costs.

Effect on Investment Decisions

- Given their limited amount, surface fees do not present any particular disadvantage to investors.

Table 15. Bonuses

How do they work?

- Bonuses are commonly paid by the investing company upon signature of an exploration and production agreement. In some cases, bonuses may be paid upon discovery, declaration of commerciality, commissioning of facilities, start of production, and/or reaching target production levels (daily or cumulative).
- Bonuses affect the project risk by increasing the exploration and development economic thresholds. To compensate for the risk, higher bonuses are balanced by lower royalties, taxes, production sharing, and/or government shares.

Advantages and Disadvantages to Host Governments

- Bonuses are easy to administer and provide an early form of revenue. The maximum level of a bonus is very much dependent on the overall fiscal terms, the characteristics of the asset, the country political risk, and the risk profile of the targeted investors.

Effect on Investment Decisions

- High signature bonuses may discourage risk-adverse investors, especially when the political risk is perceived to be high, or when there is a high level of geological uncertainty.
- Commerciality bonuses are also sensitive, as they increase the economic cut-off rate of a project.

Table 16. Government Participation

How do they work?

- Many PSCs provide an option for the host government (or the NOC) to participate in development projects. The government's participation can take several forms. Participation may be acquired as "working interest", that is, on the same terms as might be available to other joint ventures partners.[82] This may occur from the outset of a project (very rare); more often, the government may reserve the right to back into the project at some stage (normally at field development or production).[83] The right may be acquired on concessional terms or on favorable terms.[84] The most common way consists of acquiring a carried interest: the government pays for its share out of future earnings of the project. In some countries, the government backs in without repaying the investor for the expenses borne and/or the risk taken during the exploration phase. The government may exercise its rights to participate in a project directly or through a state owned enterprise.

Advantages and Disadvantages to Host Governments

- Unless non-economic reasons (to increase the sense of ownership, to facilitate transfer of technology, to increase the control over field development decisions) motivate a host government to participate directly in a project, it is not at all demonstrated that direct government participation provides benefits not otherwise available from conventional taxes.[85]

- Besides the cost and risks associated with equity participation, as well as other considerations related to the allocation of a country's resources, there may be a conflict of interest between the government as equity holder and its role as regulator overseeing the environmental and social impact of a project.

- As government participation represents a cost to investors, the higher the percentage participation, the lower other fiscal terms.

Effect on Investment Decisions

- Government participation on concessional terms reduces the cash flow and increases the risk profile of the investment.

- In case the *cash calls*[86] on governments are paid out of production, the investors are left with the burden of raising the entire financing.

- In some cases government direct participation in development activities may lead to suboptimal investment levels. Many investors regard the government participating option as a deterrent.[87]

---

[82]The working-interest owner bears the costs of exploration, development, and operation of an oil and gas asset and, in return, is entitled to a share of the production of that asset. Royalty rights, net profit interest rights, rights in production payments, and the like do not constitute working interests because they are not burdened with the responsibility to bear exploration, development and operating expenses. Likewise, contract rights to extract or share in oil and gas, or in the profits from extraction, without liability to share in the costs of production, do not constitute working interests.

[83]In the majority of cases, the contractor bears the costs and risks of exploration. If a discovery is made, the host government/NOC has the option to "back-in" for, or up to, a set percentage participating interest. Normally, the government participating interest is a working interest carried through the exploration phase.

[84]For example, exploration costs may or may not be reimbursed by the government; if reimbursed, interest that takes into account the time value of money/risk factor may or may not be applied.

[85]In some cases the risks and costs of direct participation are such that the host government would be better off solely taxing and regulating the project.

[86]To cover operating and investment costs the operator issues cash calls to each joint venture participant.

[87]Operating and technical committees' decision making processes may become less efficient. In addition, government participation reduces the percentage of reserves that can be booked by the international oil companies.

Table 17. Cost Recovery Limit[88]

How do they work?

- In many countries production sharing contracts (and sometimes concessions agreements) provide for limits on the percentage of crude oil production that can be used for cost recovery.[89] After deduction of royalties, a percentage of the remaining revenue is used to recover costs. If costs exceed the cost recovery limit, the difference is carried forward for recovery in subsequent periods.[90] Most production sharing contracts allow for unlimited carry forward. Not all costs are recoverable for the purpose of cost recovery.[91] The relevant accounting rules are generally set in the contract or in the petroleum law.

Advantages and Disadvantages to Host Governments

- The cost recovery limit ensures that in each accounting period the government will have a share of production. Cost recovery limits are less regressive than royalties. From an administrative standpoint, it is more difficult to monitor cost recovery limits than royalties.

Effect on Investment Decisions

- In PSCs that have a cost recovery limit, this would normally range between 40-60 percent (Johnston 1994). Cost recovery limits have an effect on a project's return on investment similar to a royalty. Low cost recovery limits can be quite discouraging for the development of marginal fields.

[88]Cost recovery is a concept commonly applicable to contractual arrangements.
[89]Concessionary systems normally do not have a cost recovery limit.
[90]If recoverable costs are below the cost recovery limit in some countries, excess cost oil goes directly to the government (see for example Egypt and Syria).
[91]Normally, cost recovery includes operating costs, expensed capital costs, depreciation and depletion allowance, interest on financing, investment uplift, abandonment cost fund, and unrecovered costs carried over from previous years, but there are exceptions.

Table 18. Profit Oil Split[92]

How do they work?

- In production sharing contracts profit oil (or profit gas) is the revenue that remains after deduction of royalty and cost recovery.[93] In most cases the profit oil is split according to a sliding scale defined on the basis of agreed parameters (these may include average daily production, cumulative volume of production, crude oil prices, value of production, R-Factor, and RoR).

- The profit oil (or profit gas) split between the host government and the investor is often negotiable.[94]

Advantages and Disadvantages to Host Governments

- Sliding scale profit oil splits are flexible arrangements that allow the government to provide a suitable fiscal package to a particular project without changing the overall fiscal framework.

- There appears to be a preference among governments for sliding scale profit oil based on production rates. Although these are easier to calculate than sliding scale profit oil based on R-factors or RoR, they are insensitive to changes in the price of crude oil and natural gas.

Effect on Investment Decisions

- Sliding scale profit oil split, especially if linked to R-Factors or, even better, to the return on investment, are favorably considered by investors as they lower the project specific risk by introducing flexibility in the fiscal package to suit the actual profitability of the particular project. Because of their flexibility, these types of arrangement are less likely to discourage the development of marginal fields.

[92]Profit oil applies to contractual systems.
[93]This concept corresponds to the taxable income in concessionary systems and to the service fee in service contracts. The difference is linked to the ownership of hydrocarbons (at the delivery point in production sharing contracts, at wellhead in concessionary systems. In pure service contracts all production belongs to the government).
[94]Approximately 80 percent of the profit oil split around the world has a sliding scale of some sort (D. Johnston).

Table 19. Foreign Exchange Controls

How do they work?

- Normally, investors are allowed to hold foreign exchange accounts offshore where they can earn hard currency based interest. They can receive and make payments related to the project directly offshore without the obligation to repatriate the proceeds in the country of operation. Conversion to local currency is normally limited to the amount needed to cover domestic expenditure obligations, including tax payments.

Advantages and Disadvantages to Host Governments

- There are no particular disadvantages to government that apply limited foreign exchange controls, as controls over the convertibility of currencies have become less dominant than in the past. To satisfy statistical needs, companies are normally asked to report all currency movements to the central bank. To guarantee domestic expenditure obligations, performance bonds and similar guarantees are equally effective and less costly to investors. In countries that apply strict foreign exchange regulations, petroleum contracts normally grant exemptions to oil and gas companies. This is because oil and gas normally is sold on international markets, and the proceeds of sale are often pledged as security for repayment of project loans.

Effect on Investment Decisions

- In cases where investors are required to surrender their foreign currency to the central bank at the time of export and repurchase the same at official rates to satisfy domestic project obligations, the spread between buying and selling rates (assuming no other restriction) increases the cost of doing business. Restrictive foreign currency regulations contribute to increasing the perception of sovereign risk, thus impacting a project's net present value.

Table 20. Environmental Taxes and Bonds, and Other Performance Bonds

How do they work?

- Environment protection is one area in which many governments have started to impose greater restrictions on investors' freedom to operate. In some case, investors are asked to post performance bonds as security to comply with abandonment obligations. In other cases, environmental taxes are imposed. Insurance policies to cover environmental damages are also a normal requirement. The costs associated with the protection of environment are normally considered part of the cost of operations and are tax deductible. In some cases, penalties and the cost of remedial actions associated with damages in excess of those permitted by environmental regulations are not tax deductible.

- In the petroleum sector, bonds indemnify authorities against the investors' failure to comply with contractual obligations or the terms of the concession. This safeguards the government against technical and financial failure and premature or unplanned project termination. The value of the bond is normally reduced in proportion to the value of the outstanding obligations. In some cases, holding company guarantees are used in addition to or instead of performance bonds.

Advantages and Disadvantages to Host Governments

- Direct taxation of environmental damages may be conceptually good for correcting divergence between private and social costs. However, it is quite complex to implement. In defining the tax treatment of environmental conservation and remediation, policy makers need to avoid penalizing responsible operators. This is generally done by allowing the amortization of environmental mitigation structures and equipment over their useful life and the deduction of current environmental expenses for tax calculation purposes.

- Under an ideal bonding regime, the financial risk is shifted from the government to the investor. In case of default, funds necessary to complete contractual obligations would be promptly available, which makes it possible to avoid complicated and costly legal processes.

Effect on Investment Decisions

- The tax treatment of environmental obligations is of great importance to investors. The ability to deduct the cost of environmental compliance for tax calculation purposes lowers the costs of compliance.

- Performance bonds have been greatly standardized and do not present particular problems to investors. The cost of a bond will depend on the guarantees the financial institution imposes on or is willing to accept from the investors. Those guarantees are, in turn, a function of the country and project risks and of the investor's standing and of the competitiveness of the chosen financial market.

Table 21. Local Content Obligations

How do they work?

- Local content obligations cover areas such as training requirement, local employment quotas, and the purchase of local goods and services. The objective of training obligations is to facilitate the transfer of know-how from the investor to the host government/country. A specified number of government officials are seconded to certain departments in the investor's organization for a set period of time, in some case with the dual purpose of training and oversight of operations. In addition, funds are provided each year by the investor to cover the government's training costs (these are normally between US$50–200,000 per year depending on the phase of development of the asset). Training and secondment costs are normally cost recoverable and/or tax deductible. Local employment quotas are fairly common in developing countries and are used to facilitate the creation of domestic employment. Requirements for the use of local goods and services are standard practice in developing countries.[95] Investors are normally asked to purchase the goods and services needed for their project locally if their quality is comparable with imported goods and services and their price is not higher than a percentage set in the contract (normally 10 percent).

Advantages and Disadvantages to Host Governments

- Local content obligations allow the government to achieve a diversity of policy objectives, from transfer of technology and know how to strengthening of local industries and creation of local employment. However, governments should be mindful of the need to avoid increasing inflationary pressure by allowing or imposing excessive salary scales or promoting excessive mark-ups for local goods and services. In addition, given the international nature of the oil business and the fact that oil companies generally operate in more than one country, when deciding the level of secondments, consideration should be given to the absorption capacity of the investor's organization (for example, small companies may not be able to accommodate a large number of government trainees).

Effect on Investment Decisions

- Many countries impose some form of local content obligations, and investors have developed procedures and systems to fulfill such requirements. Strict local content obligations normally increase the cost of operations and, in some cases, lower the company's efficiency. Ultimately, part of the cost is transferred to the host government through the sharing mechanism in contracts and through taxation.

---

[95]Emphasis on local content has sharply increased over the past few years. In some countries, contractual obligations have been shored up with aggressive local content legislation (see for example Angola, Equatorial Guinea, and Nigeria).

Government Take

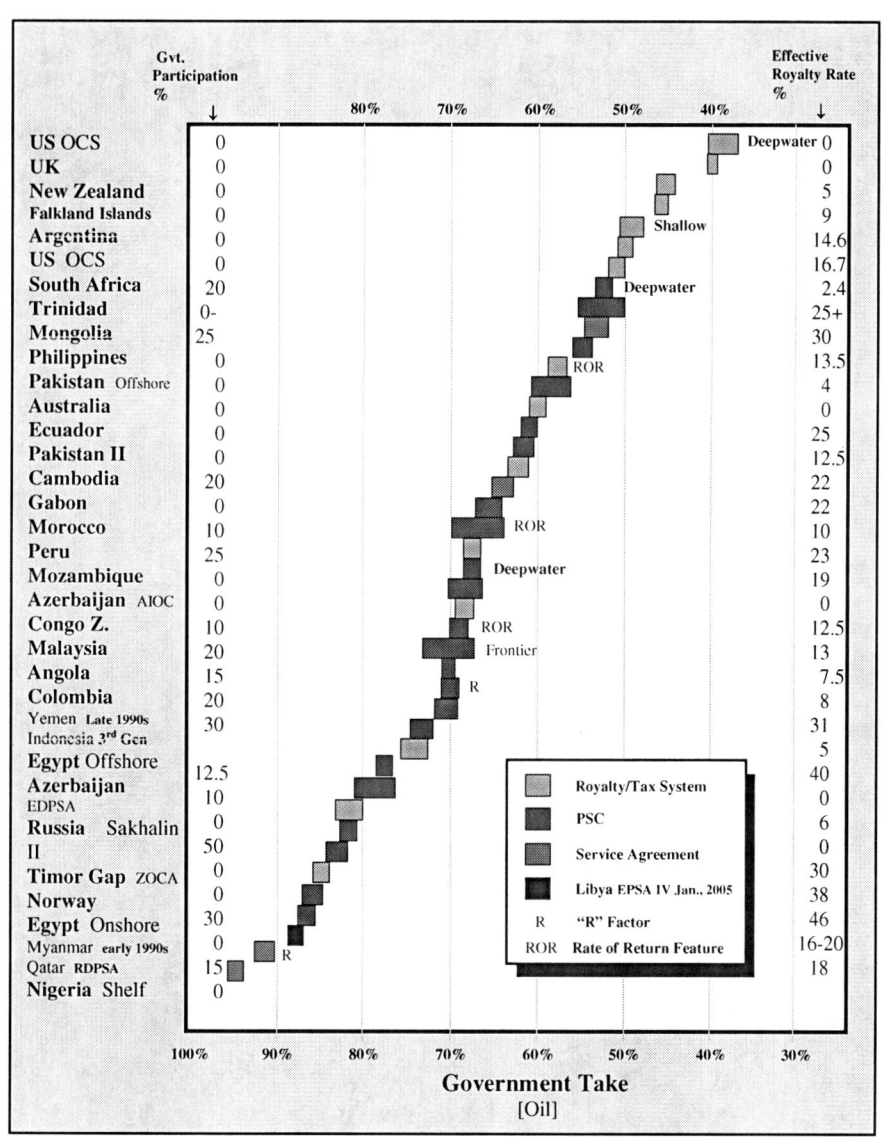

	Gvt. Participation %		Effective Royalty Rate %
US OCS	0	Deepwater 0	
UK	0		0
New Zealand	0		5
Falkland Islands	0		9
Argentina	0	Shallow	14.6
US OCS	0		16.7
South Africa	20	Deepwater	2.4
Trinidad	0-		25+
Mongolia	25		30
Philippines	0	ROR	13.5
Pakistan Offshore	0		4
Australia	0		0
Ecuador	0		25
Pakistan II	0		12.5
Cambodia	20		22
Gabon	0		22
Morocco	10	ROR	10
Peru	25		23
Mozambique	0	Deepwater	19
Azerbaijan AIOC	0		0
Congo Z.	10	ROR	12.5
Malaysia	20	Frontier	13
Angola	15		7.5
Colombia	20	R	8
Yemen Late 1990s	30		31
Indonesia 3rd Gen			5
Egypt Offshore	12.5		40
Azerbaijan EDPSA	10		0
Russia Sakhalin II	0		6
Timor Gap ZOCA	50		0
Norway	0		30
Egypt Onshore	0		38
Myanmar early 1990s	30		46
Qatar RDPSA	0	R	16-20
Nigeria Shelf	15		18
	0		

Legend:
- Royalty/Tax System
- PSC
- Service Agreement
- Libya EPSA IV Jan., 2005
- R — "R" Factor
- ROR — Rate of Return Feature

Government Take
[Oil]

Source: D. Johnston, Course Workbook, Libya Licensing Round, August 2006.

Economic Impact of
Alternative Fiscal Parameters

Table 22. Economic Impact of Alternative Fiscal Parameters—Field A

Production Sharing based on:

Field A		
Production (MBO)	20	Investor's
Price ($/Bbl)	35	
Capex ($ Million)	123	NOC's
Opex ($/Bbl)	4.54	

Base case — Investor's NPV, IRR, and Government Take

Fiscal Model	Investor's NPV ($ Million) 10.0%	12.5%	15.0%	IRR	Govt. Take (@ 10%)
No. 1 — Daily Production	70.6	50.9	35.7	26.0%	57.6%
No. 2 — Cumulative Production	70.6	50.9	35.7	26.0%	57.6%
No. 3 — R-Factor	73.5	54.7	40.0	28.2%	55.9%
No. 4 — RoR	79.9	59.6	43.9	29.0%	52.0%

Price limit ($/Bbl)

Fiscal Model	10.0%	12.5%	15.0%
No. 1	18.64	20.67	22.88
No. 2	18.64	20.67	22.88
No. 3	17.48	19.18	21.32
No. 4	17.07	18.75	20.59

Sensitivities

Fiscal Model No. 1 — Daily Production

Sensitivity		Investor's NPV ($ Million) 10.0%	12.5%	15.0%	IRR	Govt. Take (@ 10%)
Prod.	+20%	96.1	72.0	53.4	30.6%	55.7%
	-20%	45.1	29.7	18.0	20.9%	61.2%
Price	+20%	99.8	75.1	55.9	31.2%	55.4%
	-20%	41.3	26.6	15.5	20.1%	62.1%
Capex	+20%	59.5	40.2	25.5	21.9%	60.2%
	-20%	81.7	61.5	45.9	31.4%	55.5%
Opex	+20%	66.9	47.8	33.2	25.3%	58.0%
	-20%	74.3	53.9	38.3	26.7%	57.2%

Fiscal Model No. 2 — Cumulative Production

Sensitivity		Investor's NPV ($ Million) 10.0%	12.5%	15.0%	IRR	Govt. Take (@ 10%)
Prod.	+20%	96.1	72.0	53.4	30.6%	55.7%
	-20%	45.1	29.7	18.0	20.9%	61.2%
Price	+20%	99.8	75.1	55.9	31.2%	55.4%
	-20%	41.3	26.6	15.5	20.1%	62.1%
Capex	+20%	59.5	40.2	25.5	21.9%	60.2%
	-20%	81.7	61.5	45.9	31.4%	55.5%
Opex	+20%	66.9	47.8	33.2	25.3%	58.0%
	-20%	74.3	53.9	38.3	26.7%	57.2%

Fiscal Model No. 3 — R-Factor

Sensitivity		Investor's NPV ($ Million) 10.0%	12.5%	15.0%	IRR	Govt. Take (@ 10%)
Prod.	+20%	96.9	74.8	57.3	33.4%	55.3%
	-20%	49.1	33.9	22.1	22.6%	57.7%
Price	+20%	98.7	76.4	58.8	33.9%	55.9%
	-20%	46.5	31.7	20.3	22.0%	57.4%
Capex	+20%	64.9	45.7	30.9	23.8%	56.6%
	-20%	80.9	62.7	48.3	34.2%	55.9%
Opex	+20%	73.4	54.2	39.4	27.8%	54.0%
	-20%	75.3	56.4	41.6	28.8%	56.6%

Fiscal Model No. 4 — RoR

Sensitivity		Investor's NPV ($ Million) 10.0%	12.5%	15.0%	IRR	Govt. Take (@ 10%)
Prod.	+20%	100.5	77.1	58.8	33.3%	53.6%
	-20%	58.5	41.1	27.7	23.9%	49.6%
Price	+20%	104.6	80.5	61.6	34.0%	53.3%
	-20%	55.5	38.5	25.5	23.2%	49.1%
Capex	+20%	76.2	54.5	37.7	25.1%	49.0%
	-20%	85.6	66.0	50.6	34.3%	53.3%
Opex	+20%	75.6	56.1	40.9	28.2%	52.5%
	-20%	79.3	59.2	43.6	29.1%	54.3%

Fiscal terms

Fiscal Model No. 1 — Daily Production
Corp. Tax 25% | NOC 0% | C/R Limit 60%

P/O Split	Royalty
0 < DPro	3%
25 < DPro	6%
50 < DPro	8%
75 < DPro	10%
100 < DPro	12%

Fiscal Model No. 2 — Cumulative Production
Corp. Tax 25% | NOC 0% | C/R Limit 60%

P/O Split		Royalty
0 < CPro <	50	3%
50 < CPro <	150	6%
150 < CPro <	250	8%
250 < CPro <	300	10%
300 < CPro		12%

Fiscal Model No. 3 — R-Factor
Corp. Tax 25% | NOC 0% | in lieu | C/R Limit 60%

P/O Split		Royalty
R/F <	1	3%
1 < R/F <	1.5	6%
1.5 < R/F <	2	8%
2 < R/F <	2.5	10%
2.5 < R/F <	3	12%
3 < R/F		14%

Fiscal Model No. 4 — RoR
C. Tax 25% | NOC 0% | C/R Limit 60%

P/O Split			Royalty
RoR <	20%	10%	3%
20% < RoR <	25%	25%	6%
25% < RoR <	35%	40%	8%
35% < RoR <	50%	55%	10%
50% < RoR <	65%	70%	12%
65% < RoR		85%	14%

stress test

Fiscal Model	NPV(10)	NPV(15)
No. 1	5.9	(12.2)
No. 2	5.9	(12.2)
No. 3	13.9	(6.3)
No. 4	17.4	(4.4)

Table 23. Economic Impact of Alternative Fiscal Parameters—Field B

Field B — Production Sharing based on:

Production (MBO)	50 / 35
Price ($/Bbl)	
Capex ($ Million)	234
Opex ($/Bbl)	4.24

Fiscal Model No. 1 — Daily Production

	Investor's NPV ($ Million)			IRR	Govt. Take
	10.0%	12.5%	15.0%		(@ 10%)
Investor's	213.7	162.5	122.9	33.4%	54.6%
NOC's	-	-	-		
Price limit ($/Bbl)	15.05	16.54	18.22		
Sensitivities:					
Prod. +20%	278.1	215.8	167.5	38.8%	53.5%
Prod. −20%	149.4	109.2	78.3	27.5%	56.6%
Price +20%	286.8	222.9	173.4	39.4%	53.3%
Price −20%	140.7	102.0	72.4	26.7%	57.1%
Capex +20%	192.4	142.0	103.2	28.6%	56.1%
Capex −20%	235.0	183.0	142.6	39.9%	53.3%
Opex +20%	205.0	155.4	117.0	32.7%	54.9%
Opex −20%	222.4	169.6	128.8	34.2%	54.4%

Corp. Tax 60% | NOC | C/R Limit 25% | P/O Split 0%

P/O Split			Royalty
0 < D/Pro	25	30%	3%
25 < D/Pro <	50	45%	6%
50 < D/Pro <	75	60%	8%
75 < D/Pro <	100	75%	10%
100 < D/Pro		90%	12%

stress test: NPV(10) 61.8 NPV(15) 12.5

Fiscal Model No. 2 — Cumulative Production

	Investor's NPV ($ Million)			IRR	Govt. Take
	10.0%	12.5%	15.0%		(@ 10%)
Investor's	213.7	162.5	122.9	33.4%	54.6%
NOC's	-	-	-		
Price limit ($/Bbl)	15.05	16.54	18.21		
Sensitivities:					
Prod. +20%	272.0	211.8	164.8	38.8%	54.5%
Prod. −20%	149.4	109.2	78.3	27.5%	56.6%
Price +20%	286.7	222.9	173.4	39.4%	53.3%
Price −20%	140.6	102.0	72.4	26.7%	57.1%
Capex +20%	192.4	142.0	103.2	28.6%	56.1%
Capex −20%	235.0	183.0	142.6	39.9%	53.3%
Opex +20%	205.0	155.3	117.0	32.7%	54.9%
Opex −20%	222.4	169.6	128.8	34.2%	54.4%

Corp. Tax 60% | NOC | C/R Limit 25% | P/O Split 0%

P/O Split			Royalty
0 < C/Pro <	50	30%	3%
50 < C/Pro <	150	45%	6%
150 < C/Pro <	250	60%	8%
250 < C/Pro <	300	75%	10%
300 < C/Pro		90%	12%

stress test: NPV(10) 61.8 NPV(15) 12.5

Fiscal Model No. 3 — R-Factor

	Investor's NPV ($ Million)			IRR	Govt. Take
	10.0%	12.5%	15.0%		(@ 10%)
Investor's	212.4	166.1	129.4	36.6%	54.9%
NOC's	-	-	-		
Price limit ($/Bbl)	14.13	15.39	16.96		
Sensitivities:					
Prod. +20%	254.9	202.4	160.7	41.4%	54.5%
Prod. −20%	151.5	114.2	85.0	29.8%	56.6%
Price +20%	257.9	205.6	164.0	42.1%	53.3%
Price −20%	147.8	110.6	81.5	29.1%	57.1%
Capex +20%	192.5	146.4	110.2	31.0%	56.1%
Capex −20%	216.0	172.1	137.2	42.7%	53.3%
Opex +20%	203.4	158.6	123.1	35.7%	54.9%
Opex −20%	212.8	166.4	129.7	36.7%	54.4%

Corp. Tax 60% | NOC | C/R Limit 25% | P/O Split 0%

P/O Split			Royalty
R/F <	1	10%	3%
1 < R/F <	1.5	25%	6%
1.5 < R/F <	2	40%	8%
2 < R/F <	2.5	55%	10%
2.5 < R/F <	3	70%	12%
3 < R/F		85%	14%

stress test: NPV(10) 74.8 NPV(15) 23.5

Fiscal Model No. 4 — RoR

	Investor's NPV ($ Million)			IRR	Govt. Take
	10.0%	12.5%	15.0%		(@ 10%)
Investor's	211.0	164.8	128.3	36.4%	55.2%
NOC's	-	-	-		
Price limit ($/Bbl)	13.79	15.07	16.46		
Sensitivities:					
Prod. +20%	245.7	194.6	154.1	40.7%	58.9%
Prod. −20%	159.3	119.8	89.0	30.1%	53.7%
Price +20%	246.8	195.6	155.1	41.0%	59.8%
Price −20%	149.8	111.9	82.3	29.1%	54.3%
Capex +20%	205.6	155.9	117.1	31.4%	53.1%
Capex −20%	202.5	160.8	127.7	41.5%	59.8%
Opex +20%	205.3	159.8	123.8	35.7%	54.8%
Opex −20%	211.4	165.0	128.4	36.5%	56.6%

C. Tax 60% | NOC | C/R Limit 25% | P/O Split 0%

P/O Split			Royalty
RoR <		10%	3%
20% < RoR <	20%	25%	6%
25% < RoR <	25%	40%	8%
35% < RoR <	35%	55%	10%
50% < RoR <	50%	70%	12%
65% < RoR	65%	85%	14%

stress test: NPV(10) 95.6 NPV(15) 35.1

Table 24. Economic Impact of Alternative Fiscal Parameters—Field C

Field C

Production (MBO)	100
Price ($/Bbl)	35
Capex ($Million)	336
Opex ($/Bbl)	3.05

	Fiscal Model No. 1 — Daily Production					Fiscal Model No. 2 — Cumulative Production					Fiscal Model No. 3 — R-Factor					Fiscal Model No. 4 — RoR				
	Investor's NPV ($ Million)			IRR	Govt. Take (@10%)	Investor's NPV ($ Million)			IRR	Govt. Take (@10%)	Investor's NPV ($ Million)			IRR	Govt. Take (@10%)	Investor's NPV ($ Million)			IRR	Govt. Take (@10%)
Production Sharing based on:	10.0%	12.5%	15.0%			10.0%	12.5%	15.0%			10.0%	12.5%	15.0%			10.0%	12.5%	15.0%		
Investor's	515.3	408.5	325.5	49.7%	53.0%	469.7	374.9	300.6	49.4%	57.1%	456.0	373.3	306.7	55.2%	58.4%	363.4	296.8	242.9	50.4%	66.8%
NOC's	-	-	-			-	-	-			-	-	-			-	-	-		
Price limit ($/Bbl)	10.56	11.49	12.55			11.32	11.83	12.95			10.00	10.66	11.54			9.62	10.44	11.31		
Sensitivities:																				
Prod. +20%	629.3	502.4	403.7	55.7%	53.8%	578.2	465.9	377.5	56.5%	57.5%	531.5	439.0	364.2	62.6%	60.9%	374.0	309.3	256.6	55.7%	72.5%
Prod. -20%	389.1	304.1	238.2	42.0%	53.2%	361.2	283.7	223.1	41.6%	56.5%	346.4	278.4	224.1	45.5%	58.3%	339.1	270.7	216.5	44.5%	59.2%
Price +20%	660.2	528.6	426.1	57.7%	52.3%	604.7	487.8	395.8	57.7%	56.3%	529.8	438.6	364.7	63.3%	61.7%	382.1	316.3	262.6	56.5%	72.4%
Price -20%	370.3	288.4	224.9	40.7%	54.1%	334.6	262.0	205.3	40.2%	58.5%	357.0	285.3	228.4	45.0%	55.8%	327.5	260.9	208.2	43.6%	59.4%
Capex +20%	486.2	380.6	298.8	43.0%	53.7%	440.4	346.9	273.8	42.6%	53.7%	433.0	349.8	283.1	47.3%	58.8%	432.0	346.1	278.0	46.3%	58.9%
Capex -20%	544.4	436.4	352.1	58.6%	52.3%	498.9	402.9	327.3	58.6%	52.3%	447.8	370.5	307.9	64.4%	60.8%	315.8	261.8	217.7	57.5%	72.3%
Opex +20%	502.8	398.3	317.0	49.0%	53.1%	458.1	365.3	292.5	48.7%	53.1%	444.6	363.6	298.4	54.4%	58.5%	355.1	290.9	238.8	50.5%	58.5%
Opex -20%	527.7	418.7	334.0	50.3%	52.9%	481.2	384.5	308.6	50.1%	52.9%	443.5	363.4	298.9	55.3%	60.4%	357.0	292.4	239.9	50.8%	60.4%

Fiscal Model No. 1 — Corp. Tax 25%, NOC 0%, C/R Limit 60%

PVO Split		Royalty
0 < DVPro	25	30%
25 < DVPro	50	45%
50 < DVPro	75	60%
75 < DVPro	100	75%
100 < DVPro		90%

Fiscal Model No. 2 — Corp. Tax 25%, NOC 0%, C/R Limit 60%

PVO Split			Royalty
0 < CfPro <	50	30%	3%
50 < CfPro <	150	45%	6%
150 < CfPro <	250	60%	8%
250 < CfPro <	300	75%	10%
300 < CfPro		90%	12%

Fiscal Model No. 3 — Corp. Tax 25%, NOC 0%, C/R Limit 60%

PVO Split			Royalty
R/F <	1	10%	3%
1 < R/F <	1.5	25%	6%
1.5 < R/F <	2	40%	8%
2 < R/F <	2.5	55%	10%
2.5 < R/F <	3	70%	12%
3 < R/F		85%	14%

Fiscal Model No. 4 — C. Tax 25%, NOC 0%, C/R Limit 60%

PVO Split			Royalty
RoR <		10%	3%
0% < RoR <	20%	25%	6%
20% < RoR <	25%	40%	8%
25% < RoR <	35%	55%	10%
35% < RoR <	50%	70%	12%
50% < RoR <	65%	85%	14%
65% < RoR			

stress test	NPV(10)	NPV(15)	NPV(10)	NPV(15)	NPV(10)	NPV(15)	NPV(10)	NPV(15)
	(48.5)	(82.2)	211.0	111.3	242.0	137.1	247.6	139.6

Table 25. Economic Impact of Alternative Fiscal Parameters—Field D

Field D

Production Sharing based on:		
Production (MBO)	600	
Price ($/Bbl)	35	
Capex ($ Million)	4,615	
Opex ($/Bbl)	2.31	

		Fiscal Model No.1 — Daily Production					Fiscal Model No.2 — Cumulative Production					Fiscal Model No.3 — R-Factor					Fiscal Model No.4 — RoR				
		Investor's NPV ($ Million)			IRR	Govt. Take (@ 10%)	Investor's NPV ($ Million)			IRR	Govt. Take (@ 10%)	Investor's NPV ($ Million)			IRR	Govt. Take (@ 10%)	Investor's NPV ($ Million)			IRR	Govt. Take (@ 10%)
		10.0%	12.5%	15.0%			10.0%	12.5%	15.0%			10.0%	12.5%	15.0%			10.0%	12.5%	15.0%		
Investor's		386.4	(72.9)	(417.5)	12.1%	91.6%	(14.4)	(321.0)	(561.2)	9.9%	100.3%	1,751.8	1,111.7	612.2	19.2%	61.7%	2,423.0	1,623.6	1,006.6	21.2%	47.1%
NOC's		-	-	-			-	-	-			-	-	-			-	-	-		
Price limit ($/Bbl)		29.41	36.35	44.38			35.23	41.02	47.12			20.30	23.76	27.65			19.44	22.36	25.55		
Sensitivities:																					
Prod.	+20%	606.7	106.3	(269.5)	13.1%	90.3%	237.7	(99.4)	(365.4)	11.7%	96.2%	2,461.6	1,721.8	1,140.1	22.7%	60.5%	3,277.7	2,347.1	1,624.1	24.8%	47.4%
	−20%	93.6	(316.7)	(623.0)	10.5%	96.8%	(265.8)	(548.3)	(766.9)	8.1%	109.1%	1,014.3	474.8	59.8	15.4%	65.4%	1,455.4	805.5	310.3	17.0%	50.3%
Price	+20%	853.0	305.8	(106.0)	14.3%	86.5%	417.8	52.2	(237.1)	12.9%	93.4%	2,497.1	1,749.9	1,162.7	22.8%	60.4%	3,325.0	2,383.8	1,653.0	24.9%	47.3%
	−20%	(109.8)	(482.4)	(759.6)	9.4%	103.9%	(430.2)	(684.2)	(880.1)	6.8%	115.1%	974.0	443.2	34.6	15.2%	65.8%	1,402.4	764.6	278.3	16.8%	50.8%
Capex	+20%	8.6	(462.6)	(813.7)	10.0%	99.8%	(428.7)	(742.8)	(985.9)	7.4%	111.1%	1,386.9	716.5	198.9	16.2%	63.9%	1,963.6	1,147.1	524.0	17.8%	48.9%
	−20%	747.4	299.2	(38.7)	14.7%	85.9%	407.2	105.2	(134.2)	13.5%	92.3%	2,144.5	1,525.6	1,038.3	23.7%	59.6%	2,676.9	1,934.8	1,355.6	25.3%	49.6%
Opex	+20%	358.5	(93.8)	(433.3)	11.9%	92.0%	(27.5)	(331.8)	(570.3)	9.8%	100.6%	1,714.5	1,082.2	588.5	19.1%	61.9%	2,374.0	1,585.6	976.6	21.1%	47.2%
	−20%	414.4	(52.1)	(401.6)	12.2%	91.1%	(1.3)	(310.2)	(552.2)	10.0%	100.0%	1,789.2	1,141.3	635.9	19.4%	61.6%	2,472.1	1,661.7	1,036.5	21.4%	46.9%

Fiscal Model No.1

Corp. Tax	25%		C/R Limit	60%
NOC	0%			

P/O Split		D/Pro		Royalty
0	< D/Pro <	25		3%
25	< D/Pro <	50		6%
50	< D/Pro <	75		8%
75	< D/Pro <	100		10%
100	< D/Pro			12%

stress test | NPV(10) (1,012.0) NPV(15) (1,571.7)

Fiscal Model No.2

Corp. Tax	25%		C/R Limit	60%
NOC	0%			

P/O Split		C/Pro		Royalty
0	< C/Pro <	50		3%
50	< C/Pro <	150		6%
150	< C/Pro <	250		8%
250	< C/Pro <	300		10%
300	< C/Pro			12%

stress test | NPV(10) (1,092.7) NPV(15) (1,556.0)

Fiscal Model No.3

Corp. Tax	25%		C/R Limit	60%
NOC	0%			

P/O Split		R/F		Royalty
	R/F <	1		3%
1	< R/F <	1.5		6%
1.5	< R/F <	2		8%
2	< R/F <	2.5		10%
2.5	< R/F <	3		12%
3	< R/F			14%

stress test | NPV(10) (216.4) NPV(15) (973.3)

Fiscal Model No.4

C.Tax	25%		C/R Limit	60%
NOC	0%			

P/O Split		RoR		Royalty
	RoR <	5%		3%
5%	< RoR <	15%		6%
15%	< RoR <	25%		8%
25%	< RoR <	35%		10%
35%	< RoR <	45%		12%
45%	< RoR			14%

stress test | NPV(10) (87.8) NPV(15) (904.4)

Government Take and Project IRR at Different Levels of Cost Recovery Limit

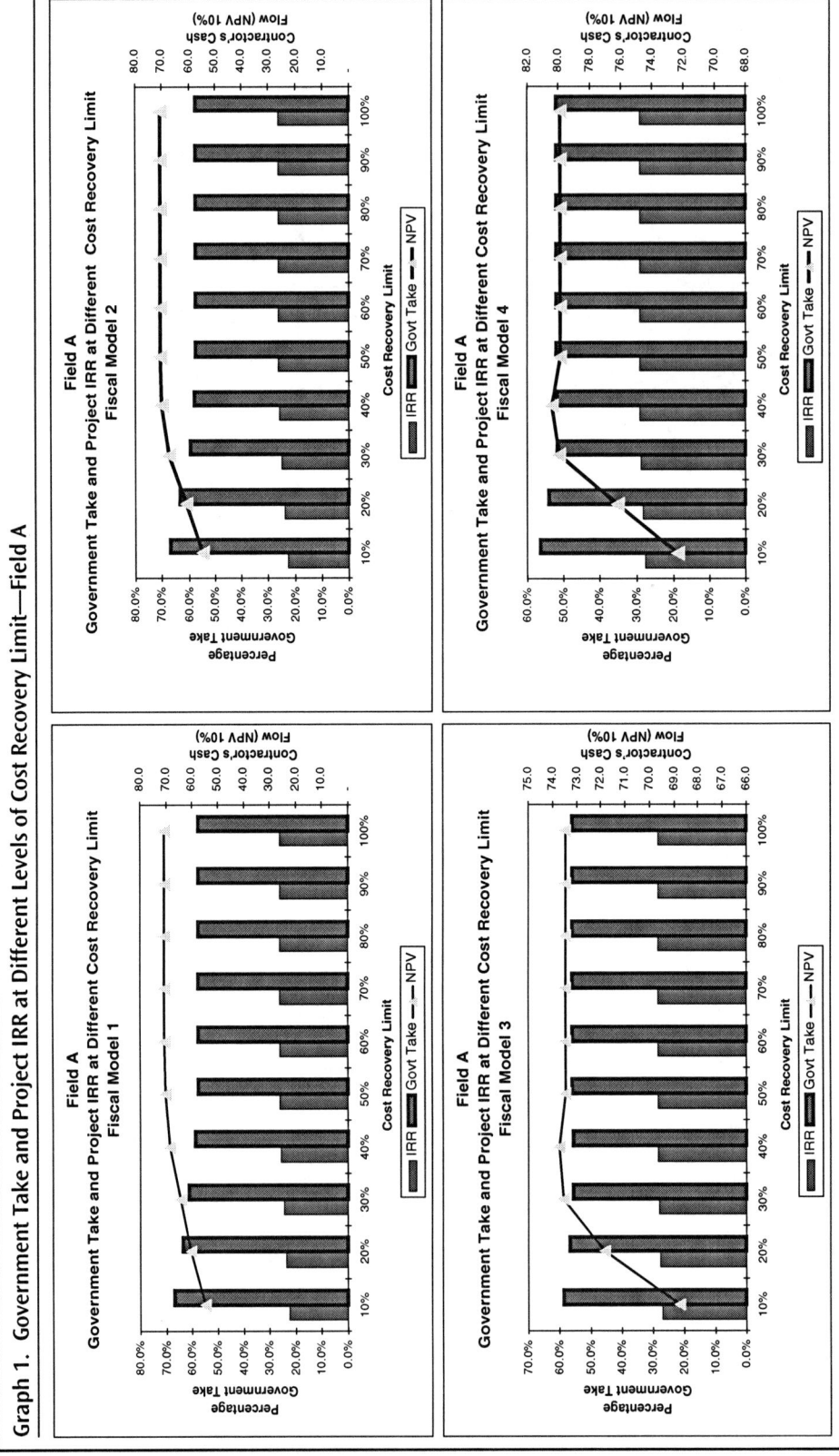

Graph 1. Government Take and Project IRR at Different Levels of Cost Recovery Limit—Field A

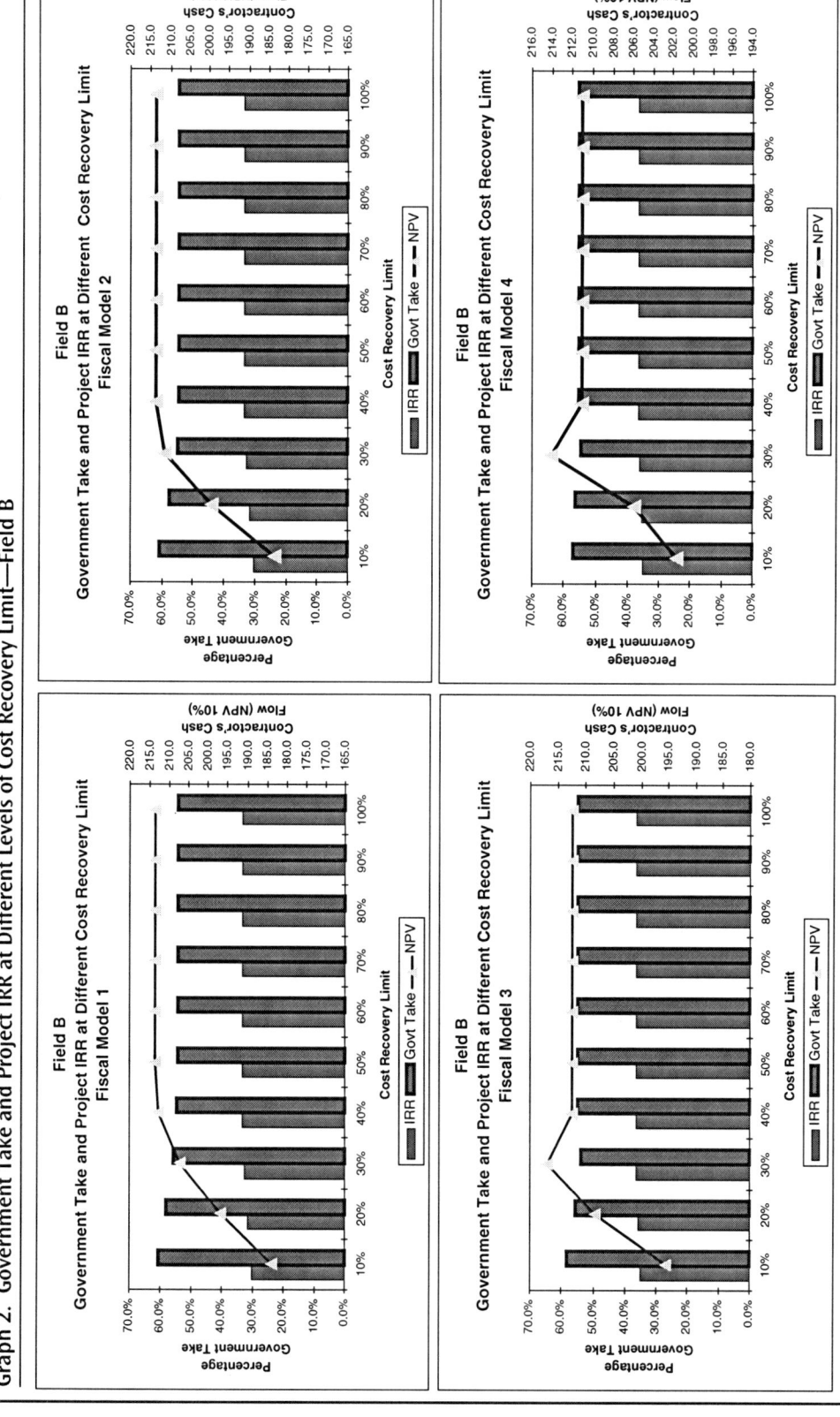

Graph 2. Government Take and Project IRR at Different Levels of Cost Recovery Limit—Field B

Graph 3. Government Take and Project IRR at Different Levels of Cost Recovery Limit—Field C

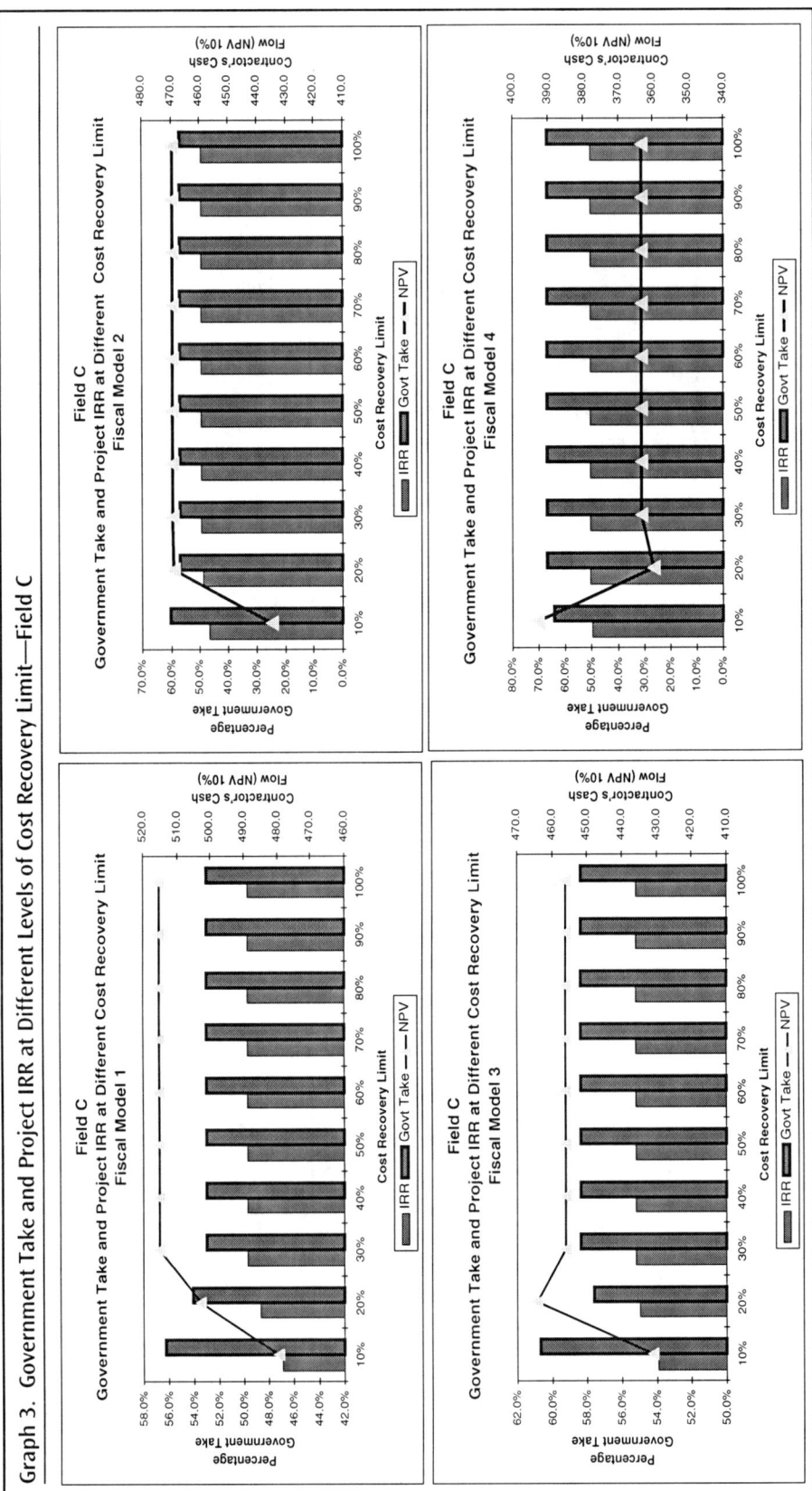

Note: The combined effect of royalties and cost recovery limit may produce a Government Take above 100 percent. In these cases, the graphs conventionally show a Government Take of 101 percent.

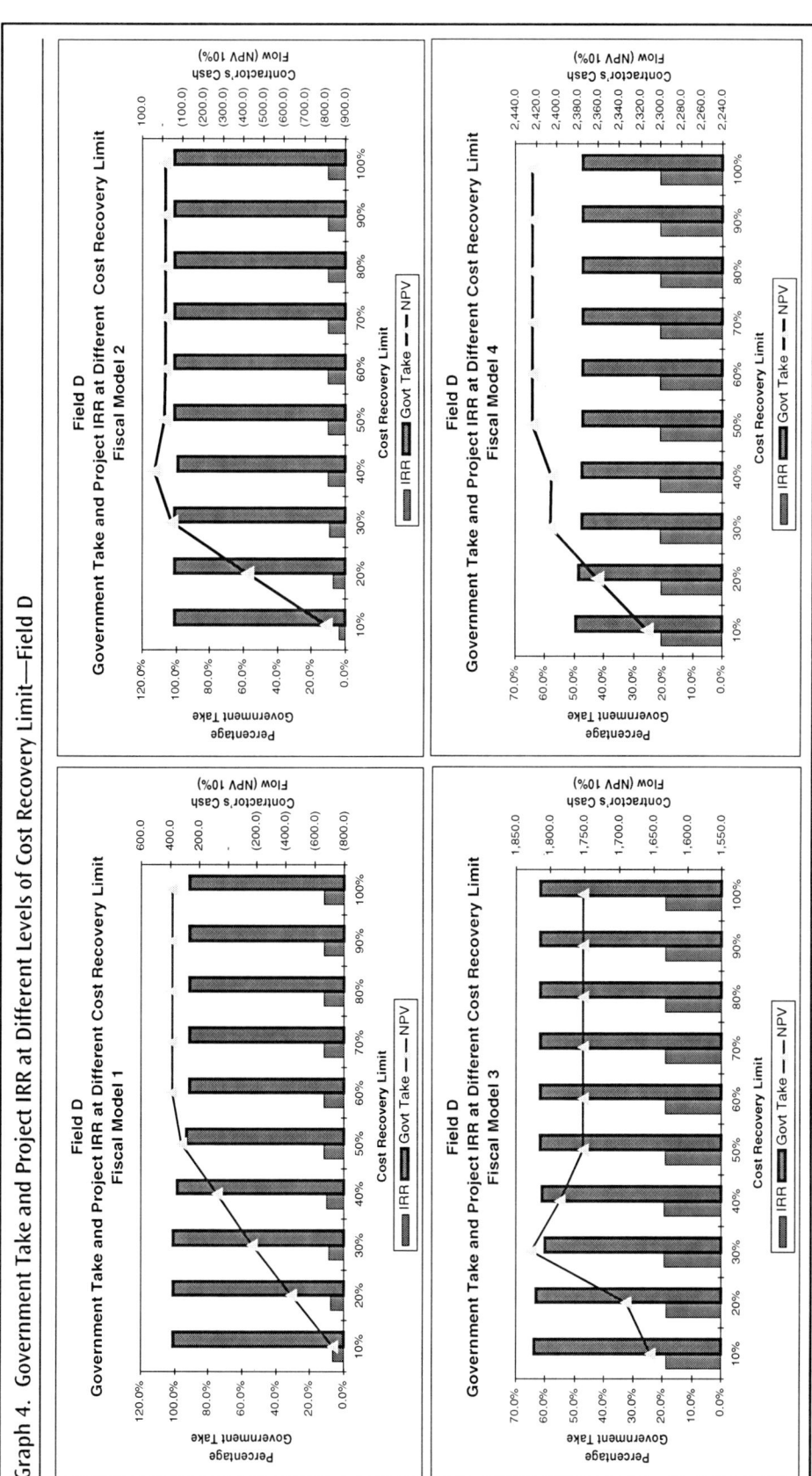

Graph 4. Government Take and Project IRR at Different Levels of Cost Recovery Limit—Field D

Note: The combined effect of royalties and cost recovery limit may produce a government take above 100 percent. In these cases, the graphs conventionally show a government take of 101 percent.

Field A, Fiscal Model 1

Alternative Triggers

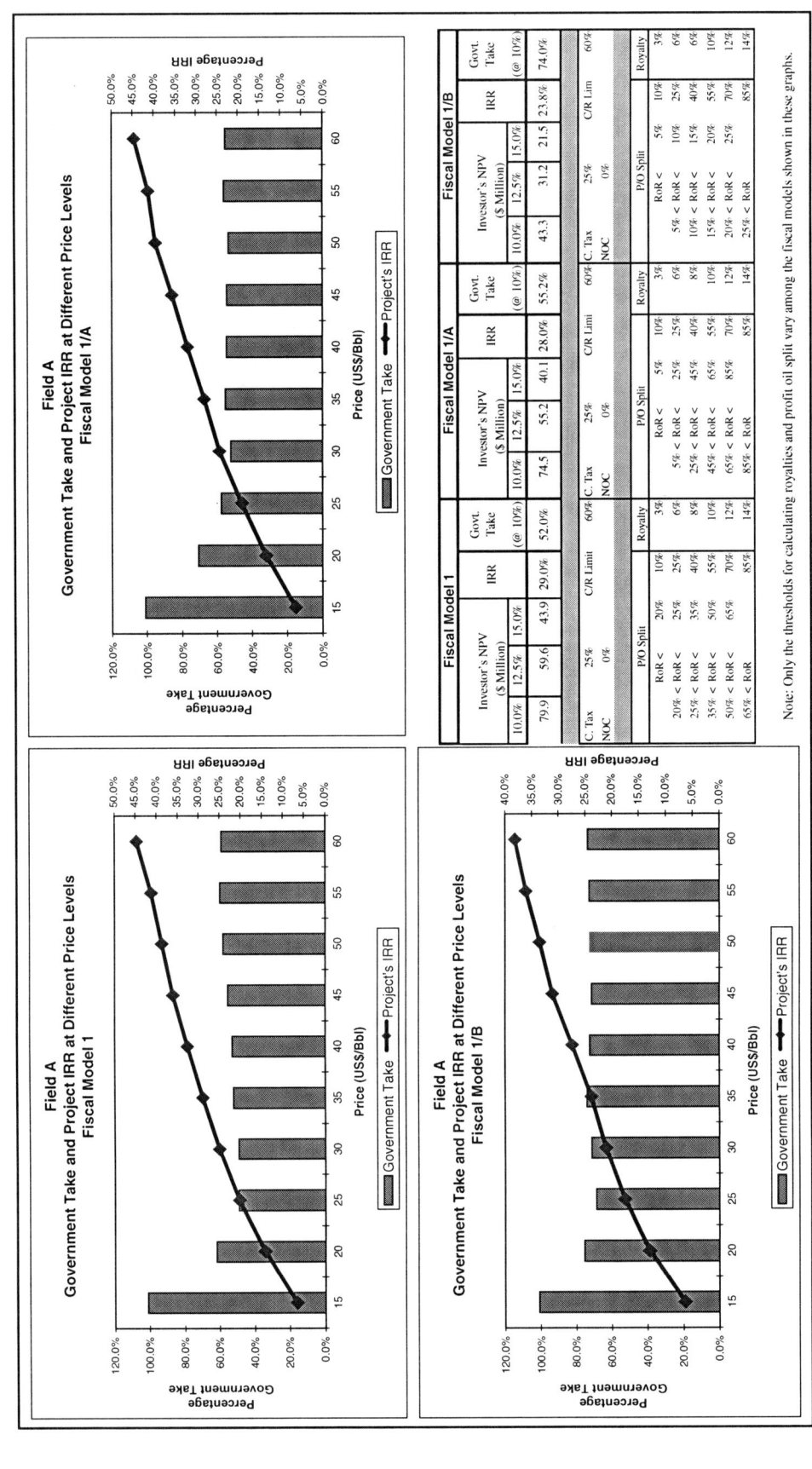

Note: Only the thresholds for calculating royalties and profit oil split vary among the fiscal models shown in these graphs.

Government Take and Project IRR at Different Price Levels

Graph 5. Government Take and Project IRR at Different Price Levels—Field A

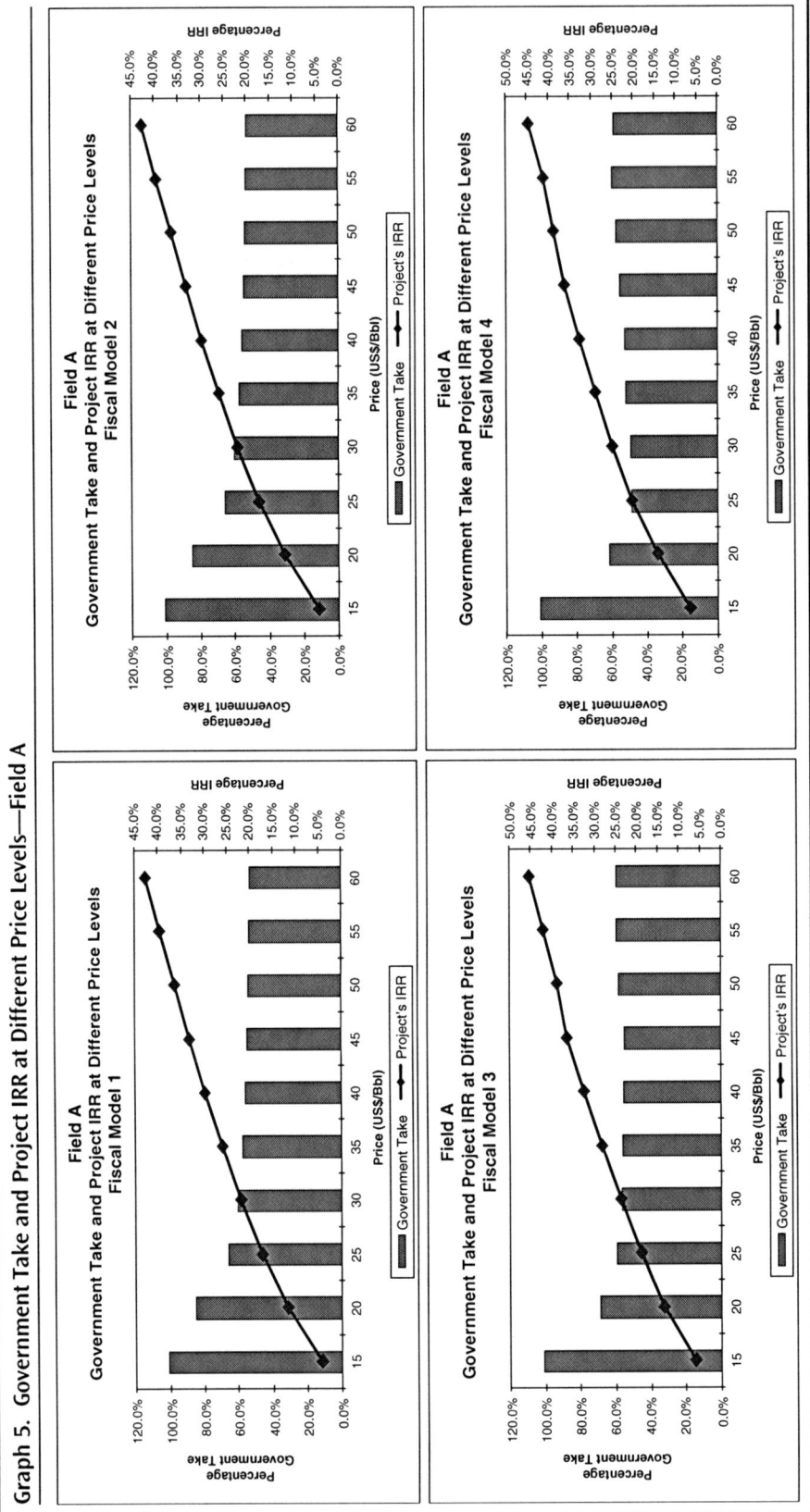

Note: The combined effect of royalties and cost recovery limit may produce a government take above 100 percent. In these cases, the graphs conventionally show a government take of 101 percent.

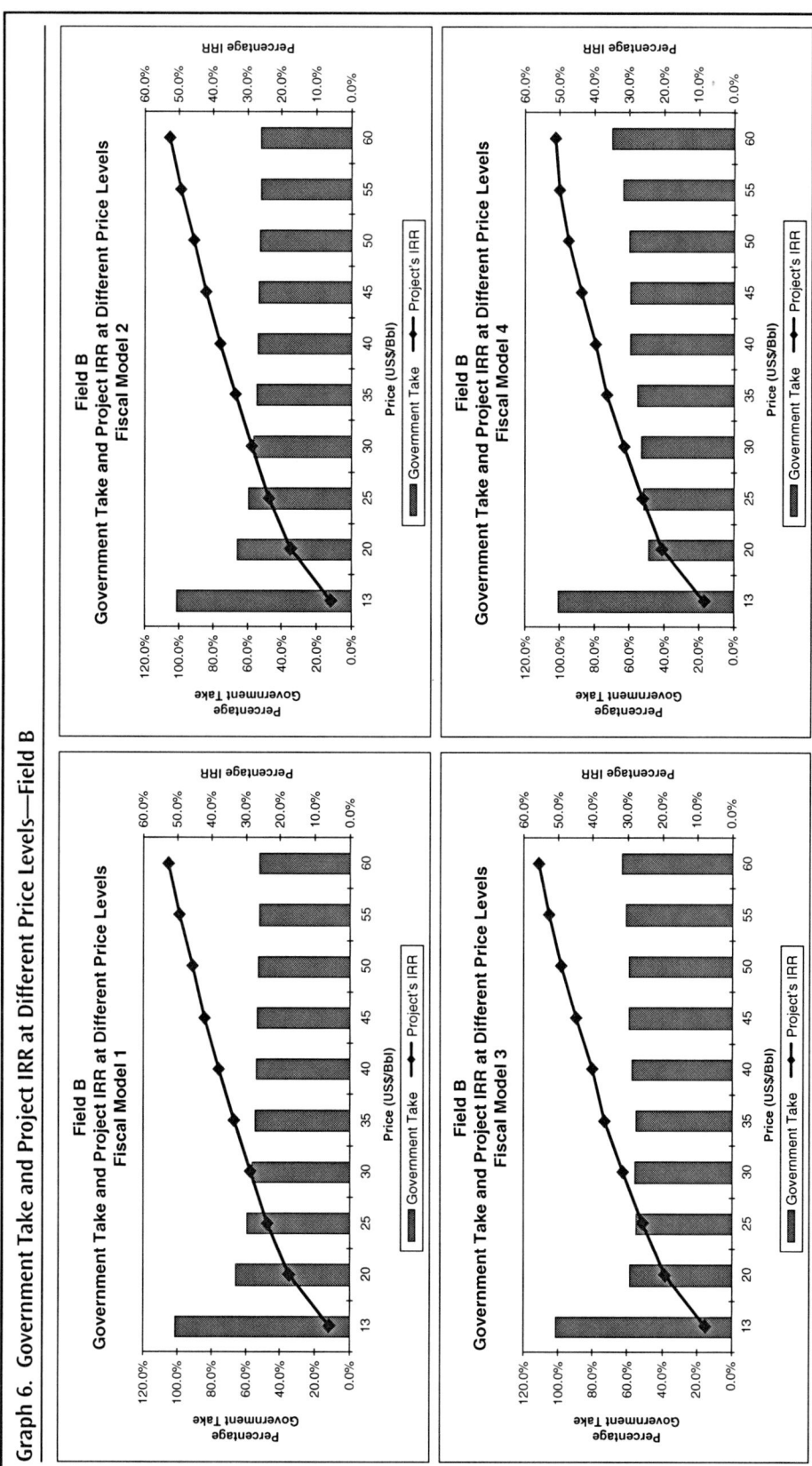

Graph 6. Government Take and Project IRR at Different Price Levels—Field B

Note: The combined effect of royalties and cost recovery limit may produce a government take above 100 percent. In these cases, the graphs conventionally show a government take of 101 percent.

Graph 7. Government Take and Project IRR at Different Price Levels—Field C

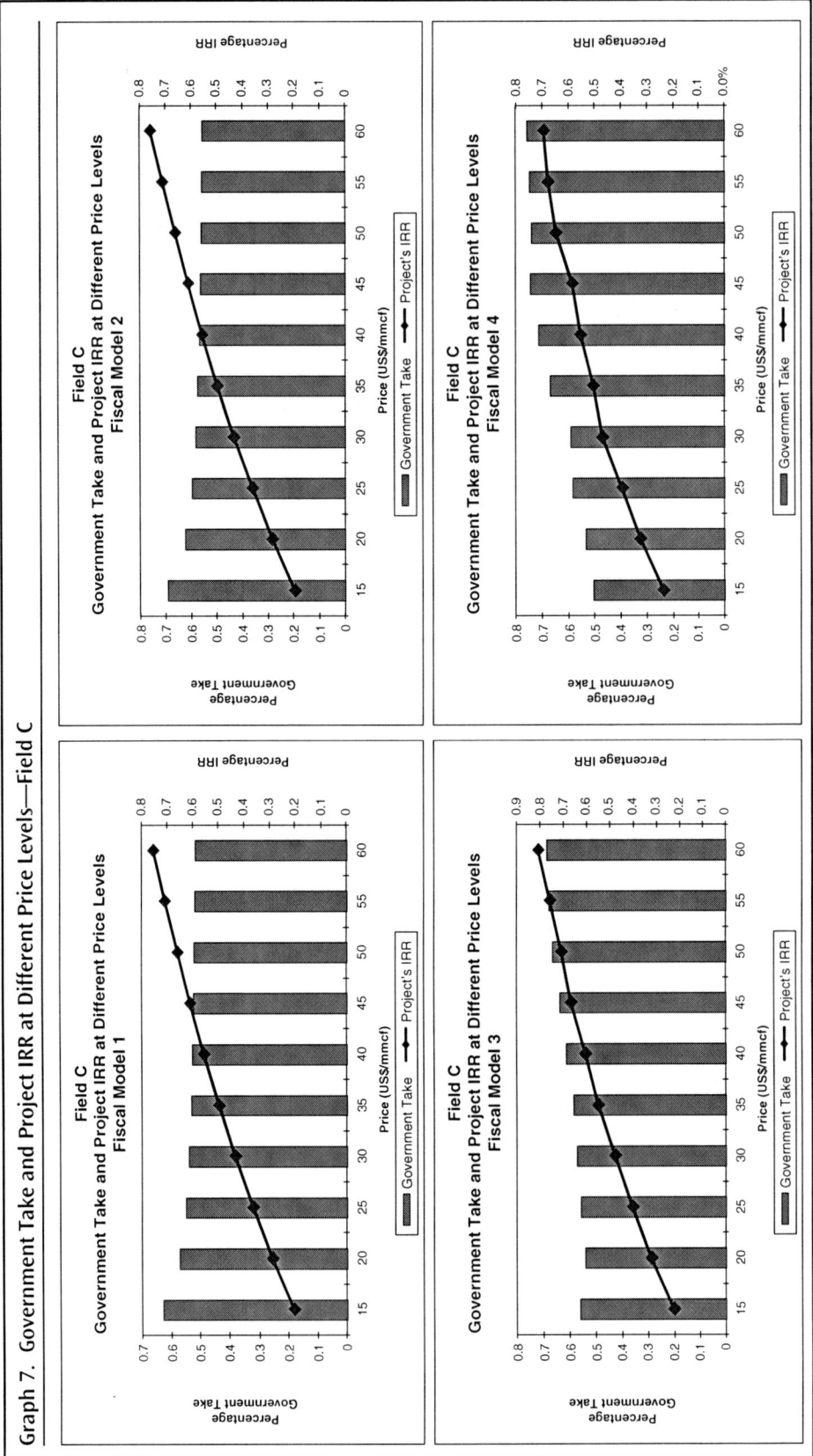

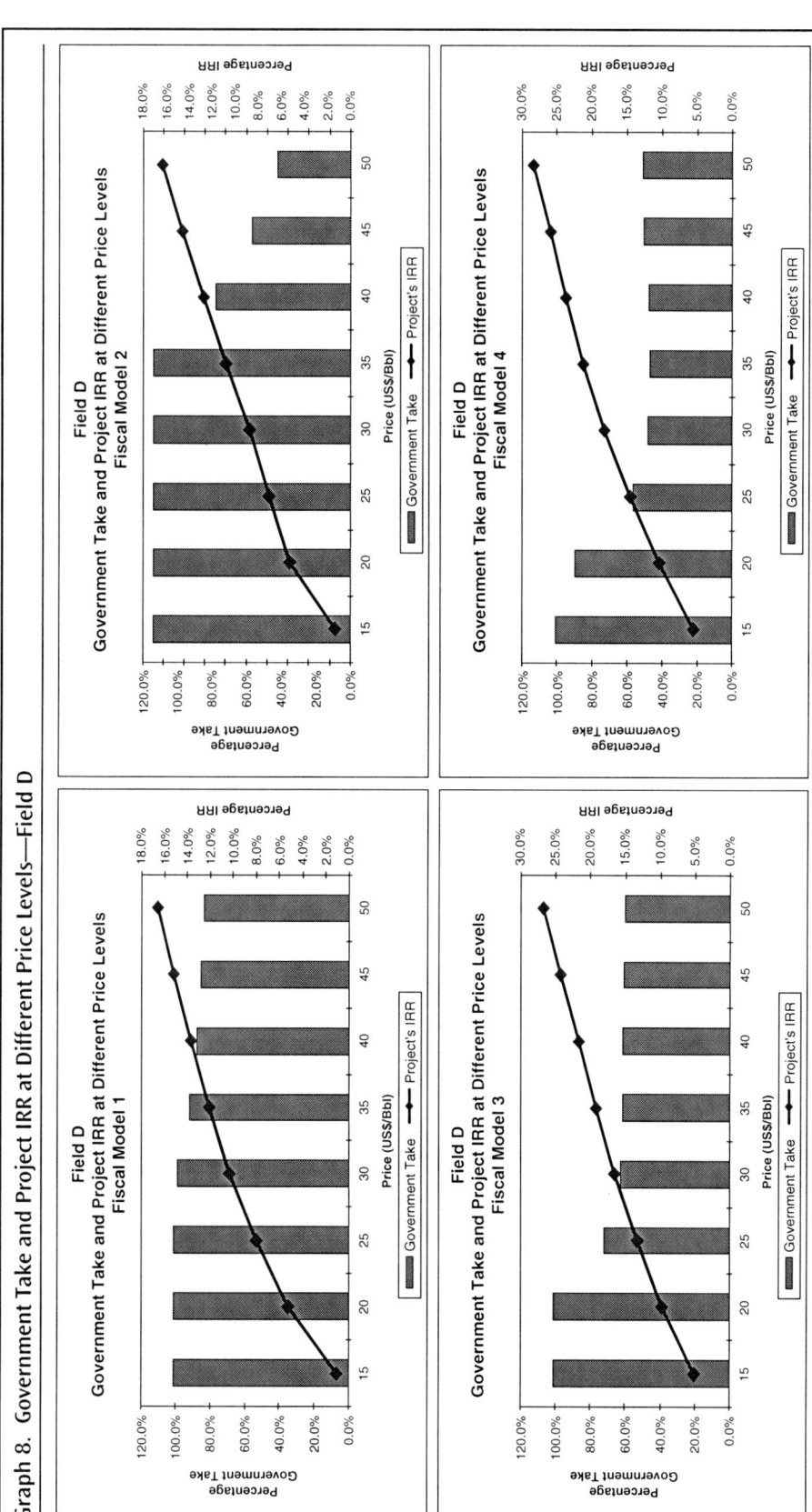

Graph 8. Government Take and Project IRR at Different Price Levels—Field D

Note: The combined effect of royalties and cost recovery limit may produce a government take above 100 percent. In these cases, the graphs conventionally show a government take of 101 percent.

Government Participating Interest

Table 26. Government Participating Interest—Field A

Production Sharing based on:

Field A		
Production (MBO):	20	Investor's
Price ($/Bbl):	35	
Capex ($ Million):	123	NOCs
Opex ($/Bbl):	4.54	
Price limit ($/Bbl):		

Fiscal Model No. 1/NOC — Daily Production

	Investor's NPV ($ Million) 10.0%	12.5%	15.0%	IRR	Govt. Take (@ 10%)
Investor's	46.7	32.5	21.7	23.3%	72.0%
NOCs	23.9	18.3	14.1		
	19.53	21.90	24.50		
Sensitivities:					
Prod. +20%	64.5	47.3	34.0	27.4%	70.2%
Prod. −20%	28.7	17.7	9.2	18.8%	75.3%
Price +20%	67.1	49.5	35.8	27.9%	70.0%
Price −20%	26.1	15.5	7.4	18.1%	76.0%
Capex +20%	38.4	24.5	13.9	19.7%	74.3%
Capex −20%	54.9	40.5	29.4	28.0%	70.1%
Opex +20%	44.1	30.4	19.9	22.7%	72.3%
Opex −20%	49.3	34.7	23.4	23.9%	71.6%

Corp. Tax: 25% / NOC: 30% — C/R Limit: 60%

P/O Split		Royalty
0 < D/Pro < 25	30%	3%
25 < D/Pro < 50	45%	6%
50 < D/Pro < 75	60%	8%
75 < D/Pro < 100	75%	10%
100 < D/Pro	90%	12%

stress test: NPV(10) = 0.7 NPV(15) = (12.8)

Fiscal Model No. 2/NOC — Cumulative Production

	Investor's NPV ($ Million) 10.0%	12.5%	15.0%	IRR	Govt. Take (@ 10%)
Investor's	46.7	32.5	21.7	23.3%	72.0%
NOCs	24	18	14		
	19.52	21.90	24.50		
Prod. +20%	64.5	47.3	34.0	27.4%	70.2%
Prod. −20%	28.7	17.7	9.2	18.8%	75.3%
Price +20%	67.1	49.5	35.8	27.9%	70.0%
Price −20%	26.1	15.5	7.4	18.1%	76.0%
Capex +20%	38.4	24.5	13.9	19.7%	74.3%
Capex −20%	54.9	40.5	29.4	28.0%	70.1%
Opex +20%	44.1	30.4	19.9	22.7%	72.3%
Opex −20%	49.3	34.7	23.4	23.9%	71.6%

Corp. Tax: 25% / NOC: 30% — C/R Limit: 60%

P/O Split		Royalty
0 < C/Pro < 50	30%	3%
50 < C/Pro < 150	45%	6%
150 < C/Pro < 250	60%	8%
250 < C/Pro < 300	75%	10%
300 < C/Pro	90%	12%

stress test: NPV(10) = 0.7 NPV(15) = (12.8)

Fiscal Model No. 3/NOC — R-Factor

	Investor's NPV ($ Million) 10.0%	12.5%	15.0%	IRR	Govt. Take (@ 10%)
Investor's	48.7	35.2	24.6	25.1%	70.8%
NOCs	24.8	19.5	15.3		
	18.24	20.27	22.98		
Prod. +20%	65.1	49.3	36.8	29.6%	69.9%
Prod. −20%	31.6	20.6	12.1	20.1%	72.8%
Price +20%	66.4	50.4	37.8	30.0%	70.3%
Price −20%	29.7	19.0	10.8	19.6%	72.7%
Capex +20%	42.2	28.4	17.6	21.2%	71.8%
Capex −20%	54.3	41.3	31.1	30.2%	70.4%
Opex +20%	48.6	34.9	24.2	24.8%	69.5%
Opex −20%	49.9	36.4	25.7	25.6%	71.2%

Corp. Tax: 25% / NOC: 30% — C/R Limit: 60% — in lieu

P/O Split		Royalty
R/F < 1	30%	3%
1 < R/F < 1.5	45%	6%
1.5 < R/F < 2	60%	8%
2 < R/F < 2.5	75%	10%
2.5 < R/F < 3	90%	12%
3 < R/F		

stress test: NPV(10) = 6.3 NPV(15) = (8.6)

Fiscal Model No. 4/NOC — RoR

	Investor's NPV ($ Million) 10.0%	12.5%	15.0%	IRR	Govt. Take (@ 10%)
Investor's	53.1	38.7	27.4	25.8%	68.1%
NOCs	26.7	21.0	16.5		
	17.80	19.76	21.91		
Prod. +20%	67.6	50.9	37.8	29.6%	68.8%
Prod. −20%	38.1	25.6	16.0	21.4%	67.2%
Price +20%	70.4	53.2	39.8	30.2%	68.5%
Price −20%	36.0	23.8	14.5	20.8%	67.0%
Capex +20%	50.1	34.5	22.4	22.5%	66.5%
Capex −20%	57.6	43.6	32.7	30.4%	68.6%
Opex +20%	50.2	36.2	25.3	25.1%	68.5%
Opex −20%	52.7	38.4	27.2	25.8%	69.6%

C. Tax: 25% / NOC: 30% — C/R Limit: 60%

P/O Split		Royalty
RoR < 20%	20%	3%
20% < RoR < 25%	25%	6%
25% < RoR < 35%	35%	8%
35% < RoR < 50%	50%	10%
50% < RoR < 65%	65%	12%
65% < RoR		14%

stress test: NPV(10) = 8.8 NPV(15) = (7.3)

(1) The % Government Take includes the NOC's share of benefits.

(2) NOC is carried throughout the exploration phase. No interest is applied to the carry.

Bibliography

Allen, F.H., and R.D. Seba. 1993. *Economics of Worldwide Petroleum Production*. Tulsa, Okla.: Oil and Gas Consultants International (OGCI), Inc.

Anderson, O.L. 1998. "Royalty valuation: Should royalty obligations be determined intrinsically, theoretically, or realistically?" *Natural Resources Law Journal*. 37:611.

Barrows, G.H. 1993. *Worldwide Concession Contracts and Petroleum Legislation*. Tulsa, Okla.: PennWell Books.

Barrows, G.H. 1994. *World Fiscal System for Oil*. Calgary, Alberta: Van Meurs & Ass. Ltd., Calgary.

Baunsgaard T. 2001. *A premier on Mineral Taxation*. IMF working paper, WP/01/139.

Boudreaux, D.O., D.R. Ward, P. Boudreaux, and S.P. Ward. 1991. "An inquiry into the capital budgeting process and analytical procedures utilized by firms in the oil and gas extraction industry." *Petroleum Accounting and Financial Management Journal* p. 24–34.

Brealey, R.A. and S.C. Myers. 1991. *Principles of Corporate Finance*. New York: McGraw-Hill.

Bunter, M.A.G. 2002. *The Promotion and Licensing of Petroleum Prospective Acreage*. The Haque: Kluiwer Law International.

Davis, J.M., Ossowski, R., Fedelino, A. 2003. *Fiscal Policy Formulation and Implementation in Oil-Producing Countries*. Washington, D.C.: International Monetary Fund.

Deloitte. 2005. Oil and Gas Survey, 2004–2005. Prepared for the Aberdeen and Grampian Chamber of Commerce. http://www.agcc.co.uk/policy

Dougherty, E.L. 1985. "Guidelines for proper application of four commonly used investment criteria." *Proceedings of the Society of Petroleum Engineers Hydrocarbon Economics and Evaluation Symposium, Dallas, TX*, SPE Paper 13770.

Ehrhardt, M.C. 1994. *The Search for Value: Measuring the Company's Cost of Capital*. Boston, Mass.: Harvard Business School Press.

Evans, D.J. 2006. *Social Discount Rates for the European Union.* Working Paper no. 2006-20. Milan: Universita' degli studi di Milano.

Gallun, R.A., C.J. Wright, L.M. Nichols, and J.W. Stevenson. 2001. *Fundamentals of Oil & Gas Accounting.* Tulsa, Okla.: PennWell Books.

Garnaut, R., and A. Clunies Ross. 1975. "Uncertainty, risk aversion and the taxing of natural resource projects." *Economic Journal* 85(2):272–87.

Gresik, Thomas A. 2001. "The Taxing Task of Taxing Transnationals." Working Paper 284.

Johnston, D. 1993. "Thinking of going international?" *Petroleum Accounting and Financial Management Journal* 13(2):84–103.

———. 1994a. "Global petroleum fiscal systems compared by contractor take." *Oil and Gas Journal* 92(50):47–50.

———. 1994b. *International Petroleum Fiscal Systems and Production Sharing Contracts.* Tulsa, Okla.: PennWell Books.

———. 2000. "Current developments in production sharing contracts and international petroleum concerns: Economic modeling/auditing: Art or science?" *Petroleum Accounting and Financial Management Journal* 19(3):120–138.

———. 2002. "Current developments in production sharing contracts and international concerns: Retrospective government take—not a perfect statistic." *Petroleum Accounting and Financial Management Journal* 21(2):101–08.

———. 2003. *International Exploration Economics, Risk, and Contract Analysis.* Tulsa, Okla.: PennWell Books.

Kaiser, M.J., and A.G. Pulsipher. 2004. *Fiscal System Analysis: Concessionary and Contractual Systems Used in Offshore Petroleum Arrangements.* Minerals Management Service, U.S. Department of the Interior.

Kemp A. 1987a. "Economic considerations in the taxation of petroleum exploitation." In K. Khan, ed., *Petroleum Resources and Development Economic, Legal and Policy Issues for Developing Countries.*

———. 1987b. *Petroleum Rent Collection Around the World.* Halifax, Nova Scotia: The Institute for Research on Public Policy.

———. 1996. *Pros and Cons of Royalty.* Oxford Energy Forum.

Kretzschmar, G.L., and P. Moles. 2006. *The Impact of Tax Shocks and Oil Price Volatility on Risk: A Study of North Sea Oilfield Projects.* W.P. 06.01, University of Edinburgh.

Kumar, R. 1991. "Taxation for a cyclical industry." *Resources Policy* No. 2:133–48.

McPherson, C.P, and K. Palmer. 1984. "New Approaches to Profit Sharing in Developing Countries." *Oil & Gas Journal,* June 25.

Mian, M.A. 2002. *Project Economics and Decision Analysis, Vol. 1: Deterministic Models.* Tulsa, Okla.: PennWell Books.

Otto, J.M. 1995. "Legal Approaches to Assessing Mineral Royalties." *Taxation of Mineral Enterprises.* Londong: Graham & Trotman.

Rapp, W.J., B.L. Litvak, G.P. Kokolis, and B. Wang. 1999. "Utilizing discounted government take analysis for comparison of international oil and gas E&P fiscal regimes." *Proceedings of the Society of Petroleum Engineers Hydrocarbon Economics and Evaluation Symposium, Dallas, TX, March 20–23.* Society of Petroleum Engineers Paper 52958.

Rutledge, I., and P. Wright. 1998. "Profitability and taxation: Analyzing the distribution of rewards between company and country." *Energy Policy* 26(10):795–812.

Seba, R.D. 1987. "The only investment selection criterion you will ever need." *Proceedings of the Society of Petroleum Engineers Hydrocarbon Economics and Evaluation Symposium, Dallas, TX, March 2–3*, SPE Paper 16310.

Smith, D. 1993. "Methodologies for comparing fiscal systems." *Petroleum Accounting and Financial Management Journal* 13(2):76–83.

———. 1987. "True government take (TGT): A measurement of fiscal terms." *Proceedings of the Society of Petroleum Engineers Hydrocarbon Economics and Evaluation Symposium, Dallas, TX, March 2–3*. SPE Paper 16308.

Stevens, P. 2003. "Resource impact: curse or blessing? A literature survey." *The Journal of Energy Literature* 9(1):3–42.

Thompson, R.S., and J.D. Wright. 1984. *Oil Property Evaluation*. Golden, Colo.: Thompson-Wright Associates.

Van Meurs, A.P. 1971. *Petroleum Economics and Offshore Mining Legislation*. Amsterdam: Elsevier Publishing Company.

Van Meurs, A.P., and A. Seck. 1995. "Governments cut takes to compete as world acreage demand falls." *Oil and Gas Journal* 93(17):78–82.

———. 1997. "Government takes decline as nations diversify terms to attract investment." *Oil and Gas Journal* 95(21):35–40.

Wood, D.A. 1990a. "Appraisal of economic performance of global exploration contracts." *Oil and Gas Journal* 88(44):48–52.

———. 1990b. "Appraisal of 20 global exploration contracts locates key variables that affect profit levels." *Oil and Gas Journal* 88(45):50–53.

———. 1993. "Economic performance of rate of return driven international petroleum production contracts." *Petroleum Accounting and Financial Management Journal* 13(2):84–103.

Eco-Audit

Environmental Benefits Statement

The World Bank is committed to preserving Endangered Forests and natural resources. We print World Bank Working Papers and Country Studies on 100 percent postconsumer recycled paper, processed chlorine free. The World Bank has formally agreed to follow the recommended standards for paper usage set by Green Press Initiative—a nonprofit program supporting publishers in using fiber that is not sourced from Endangered Forests. For more information, visit www.greenpressinitiative.org.

In 2006, the printing of these books on recycled paper saved the following:

Trees*	Solid Waste	Water	Net Greenhouse Gases	Total Energy
203	9,544	73,944	17,498	141 mil.
'40" in height and 6-8" in diameter	Pounds	Gallons	Pounds CO$_2$ Equivalent	BTUs